Dapeng Wu

Providing Quality-of-Service Guarantees in Wireless
Networks

Dapeng Wu

Providing Quality-of-Service Guarantees in Wireless Networks

Multimedia Wireless Networking

VDM Verlag Dr. Müller

Impressum/Imprint (nur für Deutschland/ only for Germany)
Bibliografische Information der Deutschen Nationalbibliothek: Die Deutsche Nationalbibliothek
verzeichnet diese Publikation in der Deutschen Nationalbibliografie; detaillierte bibliografische
Daten sind im Internet über http://dnb.d-nb.de abrufbar.
Alle in diesem Buch genannten Marken und Produktnamen unterliegen warenzeichen-, marken-
oder patentrechtlichem Schutz bzw. sind Warenzeichen oder eingetragene Warenzeichen der
jeweiligen Inhaber. Die Wiedergabe von Marken, Produktnamen, Gebrauchsnamen,
Handelsnamen, Warenbezeichnungen u.s.w. in diesem Werk berechtigt auch ohne besondere
Kennzeichnung nicht zu der Annahme, dass solche Namen im Sinne der Warenzeichen- und
Markenschutzgesetzgebung als frei zu betrachten wären und daher von jedermann benutzt
werden dürften.

Coverbild: www.purestockx.com

Verlag: VDM Verlag Dr. Müller Aktiengesellschaft & Co. KG
Dudweiler Landstr. 125 a, 66123 Saarbrücken, Deutschland
Telefon +49 681 9100-698, Telefax +49 681 9100-988, Email: info@vdm-verlag.de
Zugl.: Pittsburgh, Carnegie Mellon University, Diss., 2003

Herstellung in Deutschland:
Schaltungsdienst Lange o.H.G., Zehrensdorfer Str. 11, D-12277 Berlin
Books on Demand GmbH, Gutenbergring 53, D-22848 Norderstedt
Reha GmbH, Dudweiler Landstr. 99, D- 66123 Saarbrücken
ISBN: 978-3-639-07923-4

Imprint (only for USA, GB)
Bibliographic information published by the Deutsche Nationalbibliothek: The Deutsche
Nationalbibliothek lists this publication in the Deutsche Nationalbibliografie; detailed
bibliographic data are available in the Internet at http://dnb.d-nb.de.
Any brand names and product names mentioned in this book are subject to trademark, brand or
patent protection and are trademarks or registered trademarks of their respective holders. The use
of brand names, product names, common names, trade names, product descriptions etc. even
without
a particular marking in this works is in no way to be construed to mean that such names may be
regarded as unrestricted in respect of trademark and brand protection legislation and could thus
be used by anyone.

Cover image: www.purestockx.com

Publisher:
VDM Verlag Dr. Müller Aktiengesellschaft & Co. KG
Dudweiler Landstr. 125 a, 66123 Saarbrücken, Germany
Phone +49 681 9100-698, Fax +49 681 9100-988, Email: info@vdm-verlag.de

Copyright © 2008 VDM Verlag Dr. Müller Aktiengesellschaft & Co. KG and licensors
All rights reserved. Saarbrücken 2008

Produced in USA and UK by:
Lightning Source Inc., 1246 Heil Quaker Blvd., La Vergne, TN 37086, USA
Lightning Source UK Ltd., Chapter House, Pitfield, Kiln Farm, Milton Keynes, MK11 3LW, GB
BookSurge, 7290 B. Investment Drive, North Charleston, SC 29418, USA
ISBN: 978-3-639-07923-4

To my parents and brothers

1

ACKNOWLEDGEMENTS

First of all, I would like to express my genuine gratitude to my Ph.D. advisor, Prof. Rohit Negi, for his guidance, encouragement, and contributions in the development of my research. Without his vision, deep insight, advice, and willingness to provide funding, this work would not have been possible. His extensive knowledge, strong analytical skill, and commitment to the excellence of research and teaching are truly treasures to his students. He gives students freedom to explore the uncharted territory while providing the needed assistance at the right time. He is willing to share his knowledge and career experience and give emotional and moral encouragement. He is more than an adviser and a teacher but a role model and a friend. Working with him is proven to be an enjoyable and rewarding experience.

I would also like to thank Prof. Joseph Kabara, Prof. Srinivasan Seshan, and Prof. Ozan Tonguz for serving on my dissertation committee and providing valuable advice on my research. Special thanks are given to Prof. David O'Hallaron for supervising me and providing me with computing facilities in my early years at CMU. I am also grateful to Prof. Jon Peha for his mentoring on my research during my first year at CMU. I want to thank Prof. Tsuhan Chen for his encouragement and kind help in my professional development, and Prof. Jose Moura for enriching my knowledge in probability theory, stochastic processes, and statistical signal processing.

My fellow graduate students have made my life at CMU enjoyable and memorable. I would especially thank Arjunan Rajeswaran for his insightful comments on my research and many long hours of intellectually stimulating discussions. It was also my great pleasure to 'squander' time with him in running Marathon, playing soccer, dining at the restaurants, watching movies at the theaters, and wandering in the shopping malls. I have benefited greatly from discussing with Jin Lu on channel coding, mathematics, among others. Special thanks go to Dr. Dong Jia for his educating me with various control theories and inspiring discussions. Dr. Huimin Chen gave me many interesting lunch 'seminars' on estimation and detection theory during his visit at CMU. Qi He led me to the area of information security and deserves my thanks in particular.

Finally, this book is dedicated to my parents and brothers for their love, sacrifice and support.

2

TABLE OF CONTENTS

LIST OF TABLES

LIST OF FIGURES

14

CHAPTER 1

INTRODUCTION

The next-generation wireless networks such as the fourth generation (4G) cellular systems are targeted at supporting various applications such as voice, data, and multimedia over packet-switched networks. In these networks, person-to-person communication can be enhanced with high quality images and video, and access to information and services on public and private networks will be enhanced by higher data rates, quality of service (QoS), security measures, location-awareness, energy efficiency, and new flexible communication capabilities. These features will create new business opportunities not only for manufacturers and operators, but also for providers of content and services using these networks.

Providing QoS guarantees to various applications is an important objective in designing the next-generation wireless networks. Different applications can have very diverse QoS requirements in terms of data rates, delay bounds, and delay bound violation probabilities, among others. For example, applications such as power plant control, demand reliable and timely delivery of control commands; hence, it is critical to guarantee that no packet is lost or delayed during the packet transmission. This type of QoS guarantees is usually called *deterministic* or *hard* guarantees. On the other hand, most multimedia applications including video telephony, multimedia streaming, and Internet gaming, do not require such stringent QoS. This is because these applications can tolerate a certain small probability of QoS violation. This type of QoS guarantees is commonly referred to as *statistical* or *soft* guarantees.

For wireless networks, since the capacity of a wireless channel varies randomly with time, an attempt to provide deterministic QoS (*i.e.*, requiring zero QoS violation probability) will most likely result in extremely conservative guarantees. For example, in a Rayleigh or Ricean fading channel, the deterministically guaranteed capacity[1] (without power control) is zero! This conservative guarantee is clearly useless. For this reason, we only consider statistical QoS in this book.

To support QoS guarantees, two general approaches have been proposed. The first approach is *network-centric*. That is, the routers, switches, and base

---

[1]The capacity here is meant to be delay-limited capacity, which is the maximum rate achievable with a prescribed delay bound (see [57] for details).

stations in the network are required to provide QoS support to satisfy data rate, bounded delay, and packet loss requirements requested by applications (*e.g.*, integrated services [19, 29, 116, 163] or differentiated services [14, 68, 95]). The second approach is solely *end-system-based* and does not impose any requirements on the network. In particular, the end systems employ control techniques to maximize the application-layer quality without any QoS support from the transport network.

In this book, we address the problem of QoS provisioning primarily from the network perspective; on the other hand, we will also apply the end-system-based approach to QoS support. Next, we motivate the problem from these two perspectives.

1.1 Motivation

1.1.1 Network-centric QoS Provisioning

To provide QoS guarantees in wireless networks, a network architecture should contain the following six components: traffic specification, QoS routing, call admission control, wireless channel characterization, resource reservation, and packet scheduling.

The network architecture is illustrated in Figure 1. First, an end system uses a traffic specification procedure to specify the source traffic characteristics and desired QoS. Then, the network employs QoS routing to find path(s) between source and destination(s) that have sufficient resources to support the requested QoS. At each network node, call admission control decides whether a connection request should be accepted or rejected, based on the requested QoS, the wired link status, and/or the statistics of wireless channels. For base stations, wireless channel characterization is needed to specify the statistical QoS measure of a wireless channel, *e.g.*, a data rate, delay bound, and delay-bound violation probability triplet; this information is used by call admission control. If a connection request is accepted, resource reservation at each network node allots resources such as wireless channels, bandwidth, and buffers that are required to satisfy the QoS guarantees. During the connection life time, packet scheduling at each network node schedules packets to be transmitted according to the QoS requirements of the connections. As shown in Figure 1, in a network node, QoS routing, call admission control, resource allocation, and wireless channel characterization, are functions on the control plane, *i.e.*, performed to set up connections; packet scheduling is a function on the data plane, *i.e.*, performed to transmit packets.

The next-generation wireless networks are aimed at supporting diverse QoS requirements and traffic characteristics [60]. The success in the deployment of such networks will critically depend upon how efficiently the wireless networks

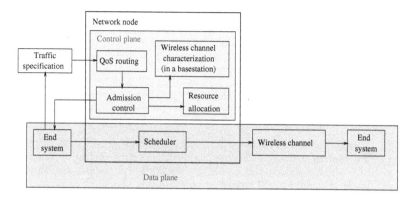

Figure 1: Network architecture for QoS provisioning.

can support traffic flows with QoS guarantees [65]. To achieve this goal, QoS provisioning mechanisms in Figure 1 (*e.g.*, admission control, resource reservation, and packet scheduling) need to be efficient and practical. For this reason, in this book, we focus primarily on designing efficient and practical admission control, resource reservation, and packet scheduling algorithms.

At a base station, to provide *connection-level QoS guarantees* such as data rate, delay and delay-violation probability, it is necessary to relate these QoS measures to the system of QoS provisioning in Figure 1. This task requires characterization of the server/service (*i.e.*, wireless channel modeling) and queueing analysis of the system. However, the existing wireless channel models (*e.g.*, Rayleigh fading model with a specified Doppler spectrum, or finite-state Markov chain models [71, 164, 165, 170]) do not explicitly characterize a wireless channel in terms of these QoS measures. To use the existing channel models for QoS support, we first need to estimate the parameters for the channel model, and then derive QoS measures from the model, using queueing analysis. This two-step approach is obviously complex [156], and may lead to inaccuracies due to possible approximations in channel modeling and deriving QoS metrics from the models.

Due to the complexity and difficulty in analyzing queues using these channel models, existing QoS provisioning schemes [4, 9, 87, 94, 106] do not utilize these channel models in their designs. This results in an undesirable situation that these schemes do not provide explicit QoS guarantees [157]. In other words, it is not clear how to characterize the QoS provisioning system under a given fading channel, in terms of connection-level QoS measures, or how to relate the control parameters of the QoS provisioning system to the QoS measures.

These observations motivate us to investigate the following problems:

1. (Wireless channel modeling) Is it possible to construct a *simple* and *accurate* wireless channel model so that connection-level QoS measures of a system can be easily obtained?

2. (QoS provisioning) If such a channel model is established, is it easy to design an *efficient* and *practical* QoS provisioning system that provides explicit QoS guarantees?

Recognizing that the key difficulty in explicit QoS provisioning, is the lack of a wireless channel model that can easily relate the control parameters of a QoS provisioning system to the QoS measures, we propose and develop a link-layer channel model termed the *effective capacity* (EC) model.

The effective capacity model captures the effect of channel fading on the queueing behavior of the link, using a computationally simple yet accurate model, and thus, is a critical tool for designing efficient QoS provisioning mechanisms. We call such an approach to channel modeling and QoS provisioning, *effective capacity approach*. Our contribution in developing the effective capacity approach are summarized in Section 1.2.

1.1.2 End-system-based QoS Support

As mentioned before, the end-system-based QoS support does not impose any requirements on the network. Such an approach is of particular significance since it does not require the participation of the networks in QoS provisioning and is applicable to both the current and future wireless Internet.[2] Note that the current wireless Internet does not provide any QoS guarantee.

End-system-based QoS support is typically application specific. In this book, we choose real-time wireless video communication as the target application. Compared with wired links, wireless channels are typically much more noisy and have both small-scale and large-scale fades [119], making the bit error rate very high; the resulting bit errors could have devastating effects on the video presentation quality. Thus, robust transmission of real-time video over wireless channels is a critical problem; and this is the problem we want to investigate in this book.

To address the above problem, we introduce adaptive QoS control for video communication over wireless channels. The objective of our adaptive QoS control is to achieve good perceptual quality and utilize network resources efficiently through adaptation to time-varying wireless channels. Our adaptive QoS control consists of optimal mode selection and delay-constrained hybrid

---

[2]By wireless Internet, we mean an IP-based network with wireless segments.

automatic repeat request (ARQ). Optimal mode selection provides QoS support (*i.e.*, error resilience) on the compression layer while delay-constrained hybrid ARQ provides QoS support (*i.e.*, bounded delay and reliability) on the link layer. Our contributions in robust wireless video transmission are summarized in the next section.

1.2   Contributions of this Book

The contributions of this book are the following.

In the area of network-centric QoS Provisioning:

- We propose and develop a link-layer channel model termed the *effective capacity* (EC) model. In this model, we characterize a wireless link by two EC functions, namely, the probability of non-empty buffer, and the QoS exponent of the connection; and we propose a simple and efficient algorithm to estimate these EC functions. We provide important insights about the relations between the EC channel model and the physical-layer channel. The key advantages of the EC link-layer modeling and estimation are (1) ease of translation into QoS measures, such as delay-bound violation probability, (2) simplicity of implementation, (3) accuracy, and hence, efficiency in admission control, and (4) flexibility in allocating bandwidth and delay for connections. In addition, the EC channel model provides a general framework, under which physical-layer fading channels such as AWGN, Rayleigh fading, and Ricean fading channels can be studied.

- With the effective capacity approach, we obtain link-layer QoS measures for various scenarios: flat-fading channels, frequency-selective fading channels, multi-link wireless networks, variable-bit-rate sources, packetized traffic, and wireless channels with non-negligible propagation delay. The QoS measures obtained can be easily translated into traffic envelope and service curve characterizations, which are popular in wired networks, such as ATM and IP, to provide guaranteed services.

- To provide QoS guarantees for K users sharing a downlink time-slotted fading channel, we develop simple and efficient schemes for admission control, resource allocation, and scheduling, which can yield substantial capacity gain. The efficiency is achieved by virtue of recently identified *multiuser diversity*. A unique feature of our work is *explicit* provisioning of statistical QoS, which is characterized by a data rate, delay bound, and delay-bound violation probability triplet. The results show that compared with a fixed-slot assignment scheme, our approach can substantially increase the statistical delay-constrained capacity of a fading channel (*i.e.*, the maximum data rate achievable with the delay-bound violation probability satisfied), when delay requirements are not very tight, while yet guaranteeing QoS at any delay requirement. For example, in the case of low signal-to-noise-ratio (SNR) and ergodic Rayleigh fading, our scheme can achieve approximately $\sum_{k=1}^{K} \frac{1}{k}$ gain for K users with loose-delay requirements, as expected from the classic paper [72] on multiuser diversity. But more importantly, when the delay bound is not loose, so that simple-minded multiuser-diversity scheduling does not directly apply, our scheme can achieve a capacity gain, and yet meet the QoS requirements.

- We design simple and efficient admission control, resource allocation, and scheduling algorithms, to provide QoS guarantees for multiple users sharing multiple parallel time-varying channels. Our scheduling algorithms consists of two sets, namely, (what we call) joint K&H/RR scheduling and Reference Channel (RC) scheduling. The joint K&H/RR scheduling, composed of K&H scheduling and Round Robin (RR) scheduling, utilizes both multiuser diversity and frequency diversity to achieve capacity gain, and the RC scheduling minimizes the channel usage while satisfying users' QoS constraints. The relation between the joint K&H/RR scheduling and the RC scheduling is that 1) if the admission control allocates channel resources to the RR scheduling due to tight delay requirements, then the RC scheduler can be used to minimize channel usage; 2) if the admission control allocates channel resources to the K&H scheduling only, due to loose delay requirements, then there is no need to use the RC scheduler. In designing the RC scheduler, we propose a *reference channel* approach and formulate the scheduler as a linear program, dispensing with complex dynamic programming approaches, by the use of a resource allocation scheme. An advantage of this formulation is that the desired QoS constraints can be *explicitly* enforced, by allotting sufficient channel resources to users, during call admission.

- We study some of the practical aspects of the effective capacity approach, namely, the effect of modulation and channel coding, and robustness against non-stationary channel gain processes. We show how to quantify the effect of practical modulation and channel coding on the effective capacity approach to QoS provisioning. We identify an important trade-off between power control and time diversity in QoS provisioning over fading channels, and propose a novel time-diversity dependent power control scheme to leverage time diversity. Compared to the power control used in the 3G wireless systems, our time-diversity dependent power control can achieve substantial capacity gains. Equipped with the time-diversity dependent power control and the effective capacity approach, we design a power control and scheduling mechanism, which is robust in QoS provisioning against large scale fading and non-stationary small scale fading.

In the area of end-system-based QoS support:

- We introduce adaptive QoS control for real-time video communication over wireless channels. The adaptive QoS control consists of optimal mode selection and delay-constrained hybrid ARQ. Specifically, optimal mode selection provides QoS support (*i.e.*, error resilience) on the compression layer while delay-constrained hybrid ARQ provides QoS support (*i.e.*, bounded delay and reliability) on the link layer.

To provide QoS on the compression layer, we employ a novel approach to rate-distortion optimized mode selection for real-time video communication over wireless channels. Different from the classical approach that only takes quantization distortion into account, our optimal mode selection also considers wireless channel characteristics and receiver behavior. To provide QoS on the link layer, we employ delay-constrained hybrid ARQ, which integrates rate-compatible punctured convolutional (RCPC) codes with delay-constrained retransmissions. Our delay-constrained hybrid ARQ has the advantage of providing reliability for the compression layer while guaranteeing delay bound and achieving high throughput.

By combining the best features of error-resilient source encoding, forward error correction (FEC) and delay-constrained retransmission, our proposed adaptive QoS control system is capable of achieving bounded delay, high reliability, and efficiency. Simulation results show that the proposed adaptive QoS control achieves better quality for real-time video under varying wireless channel conditions and utilizes network resources efficiently.

1.3 Outline of the Book

In this introduction, we have provided the motivation of the problems we address in this book: 1) network support for explicit QoS guarantees, and 2) robust transmission of real-time video over wireless channels. The remainder of this book is organized as follows.

In Chapters 2 and 3, we overview the techniques in network-centric QoS provisioning and end-system-based QoS support, respectively.

In Chapter 4, we present the technique of effective capacity channel modeling. We first motivate the link-layer channel modeling, and present an EC link-layer model for flat fading channels, which consists of two EC functions, namely, the probability of non-empty buffer, and the QoS exponent of a connection. Then, we propose a simple and efficient algorithm to estimate these EC functions. Further, we extend the EC modeling technique to frequency selective fading channels. With the EC model obtained, we derive link-layer QoS measures for multi-link wireless networks, variable-bit-rate sources, packetized traffic, and wireless channels with non-negligible propagation delay.

Armed with the new channel model, we investigate its use in designing admission control, resource reservation, and scheduling algorithms, for efficient support of QoS guarantees, in Chapters 5 and 6. Specifically, Chapter 5 addresses the problem of QoS provisioning for multiple users sharing a single channel, while Chapter 6 deals with the problem of QoS provisioning for multiple users sharing multiple parallel channels.

In Chapter 7, we investigate practical aspects of the effective capacity approach. We first discuss the effect of modulation and channel coding. We then identify the trade-off between power control and time diversity, and propose a time-diversity dependent power control scheme to leverage time diversity. Based on the time-diversity dependent power control, we design our power control and scheduling scheme, which is robust in QoS provisioning against large scale fading and non-stationary small scale fading.

In Chapter 8, we introduce adaptive QoS control for real-time video communication over wireless channels. We first sketch the overall transport architecture for real-time video communication over the wireless channel. Then, we describe our optimal mode selection algorithm and present the delay-constrained hybrid ARQ.

In Chapter 9, we summarize the book and point out future research directions.

CHAPTER 2

NETWORK-CENTRIC QOS

PROVISIONING: AN OVERVIEW

In this chapter, we overview the issues and techniques in network-centric QoS provisioning for wireless networks. We organize this chapter as follows. Section 2.1 presents various network services models. Understanding network service models and associated QoS guarantees is the first step in designing QoS provisioning mechanisms. In Section 2.2, we overview widely-used traffic models. Section 2.3 surveys packet scheduling schemes for wireless transmission. In Section 2.4, we discuss the issue of call admission control in wireless networks. Section 2.5 addresses wireless channel modeling, which plays an important role in QoS provisioning.

2.1 Network Services Models

2.1.1 The Integrated Services Model of the IETF

To support applications with diverse QoS guarantees in IP networks, the IETF Integrated Services (IntServ) Working Group has specified three types of services, namely, the *guaranteed service* [116], the *controlled-load service* [145], and the *best-effort service*.

The guaranteed service (GS) guarantees that packets will arrive within the guaranteed delivery time, and will not be discarded due to buffer overflows, provided that the flow's traffic conforms to its specified traffic parameters [116]. This service is intended for applications which need a hard guarantee that a packet will arrive no later than a certain time after it was transmitted by its sender. That is, the GS does not control the minimal or average delay of a packet; it merely controls the maximal queueing delay. Examples that have hard real-time requirements and require guaranteed service include certain audio and video applications which have fixed playback rates. Delay typically consists of two components, namely, fixed delay and queueing delay. The fixed delay is a property of the chosen path, which is not determined by the guaranteed service, but rather, by the setup mechanism. Only queueing delay is determined by the GS.

25

The controlled-load (CL) service is intended to support a broad class of applications which have been developed for use in today's Internet, but are sensitive to heavy load conditions [145]. Important members of this class are the adaptive real-time applications (e.g., *vat* and *vic*) which are offered by a number of vendors and researchers [69]. These applications have been shown to work well over lightly-loaded Internet environment, but to degrade quickly under heavy load conditions. The controlled-load service does not specify any target QoS parameters. Instead, acceptance of a request for controlled-load service is defined to imply a commitment by the network to provide the requester with a service closely approximating the QoS the same flow would receive under lightly-loaded conditions.

Both the guaranteed service and the controlled-load service are designed to support real-time applications which need different levels of QoS guarantee from the network.

The best-effort (BE) service class offers the same type of service under the current Internet architecture. That is, the network makes effort to deliver data packets but makes no guarantees. This works well for non-real-time applications which can use an end-to-end retransmission strategy (i.e., TCP) to make sure that all packets are delivered correctly. These include most popular applications like Telnet, FTP, email, Web browsing, and so on. All of these applications can work without guarantees of timely delivery of data. Another term for such non-real-time applications is *elastic*, since they are able to stretch gracefully in the face of increased delay. Note that these applications can benefit from shorter-length delays but that they do not become unusable as delays increase.

2.1.2 The Differentiated Services Model of the IETF

The implementation of the IntServ models suffers severe scalability problem. To mitigate it, the IETF specifies the Differentiated Services (DiffServ) framework for the next generation Internet [14, 96]. The DiffServ architecture offers a framework within which service providers can offer each customer a range of network services differentiated on the basis of performance. Once properly designed, a DiffServ architecture can offer great flexibility and scalability, as well as meeting the service requirements for multimedia streaming applications. The IETF DiffServ working group has specified the Assured Forwarding (AF) per hop behavior (PHB) [58]. The AF PHB is intended to provide different levels of forwarding assurances for IP packets at a node and therefore, can be used to implement multiple priority service classes.

2.1.3 The Services Model of the ATM Forum

For ATM networks, the ATM Forum [5] defines the following services: constant bit rate (CBR), real-time variable bit rate (rt-VBR), non-real-time VBR (nrt-VBR), available bit rate (ABR), and unspecified bit rate (UBR).

Under the CBR service, traffic is specified by its peak cell rate (PCR) and its associated cell delay variation (CDV) tolerance; the connection is serviced at its peak rate at each network node. Under the VBR service, a connection is characterized by PCR, sustainable cell rate (SCR), and the maximum burst size (MBS). The rt-VBR service supports slightly bursty, isochronous streams such as packet voice and video, while the nrt-VBR service is suitable for interactive streams, which are asynchronous, but still delay sensitive. Under the ABR service, the end system transmits its packets at an (instantaneous) rate dynamically set by the network so as to avoid network congestion. Under the UBR service, a connection does not declare traffic parameters and receives no QoS guarantees.

2.1.4 The Services Model for Wireless Networks

Providing QoS guarantees such as data rate, delay, and loss rate is one of the main features of the next-generation wireless networks. As we mentioned in Chapter 1, these QoS guarantees can be either deterministic or statistical. However, due to the severely conservative nature of deterministic guarantees, we only consider statistical QoS guarantees in this book.

In order to support the QoS requested by applications, network designers need to decide what kind of network services should be provided. According to the nature of wireless networks and the QoS guarantees offered, we classify network services into three categories: statistical QoS-assured service, adaptive service, and best-effort. Under statistical QoS-assured services, statistical QoS guarantees are explicitly provisioned; we will formally define statistical QoS guarantees in Chapter 4. Under best-effort services, no QoS guarantees are supported.

Adaptive services provide mechanisms to adapt traffic streams during periods of QoS fluctuations and hand-offs [152]. Adaptive services have been demonstrated to be able to effectively mitigate fluctuations of resource availability in wireless networks [6]. There have been many proposals on adaptive approaches and services in the literature, which include an "adaptive reserved service" framework [78], a wireless adaptive mobile information system (WAMIS) [3], an adaptive service based on QoS bounds and revenue [86], an adaptive framework targeted at end-to-end QoS provisioning [93], a utility-fair adaptive service [12], a framework for soft QoS control [110], a teleservice model based on an adaptive QoS paradigm [63], an adaptive QoS framework called AQuaFWiN [132], and an adaptive QoS management architecture [75], among others. Although adaptive services provide a service better than best effort, no explicit QoS guarantees is enforced.

2.2 Traffic Modeling

Traffic modeling plays an important role in QoS provisioning. It facilitates traffic specifications and accurate call admission control. Without a traffic model or characterization, measurement-based admission control needs to be employed with reduced accuracy and efficiency, compared to traffic-specification based admission control.

Traffic models fall into two categories: CBR and VBR as shown in Figure 2. For VBR, the traffic models can be either deterministic or stochastic.

The most commonly used deterministic model is linear bounded arrival process (LBAP) [34]. It has two parameters: token generating rate ρ and token bucket size σ. The amount of source traffic over any time interval of length τ is upper bounded by $\rho\tau + \sigma$, $i.e.$,

$$A(t, t + \tau) \leq \rho\tau + \sigma, \tag{1}$$

where $A(t, t + \tau)$ is the amount of source traffic generated during $[t, t + \tau)$. The simplicity of the LBAP model makes it very useful for traffic shaping and policing, which are required to ensure that the source traffic conforms to the declared characterization. Actually, the (σ, ρ)-leaky bucket regulator [131], which is the most widely used traffic shaping and policing scheme, produces traffic that can be characterized by the LBAP model. Hence, the LBAP is also called *leaky-bucket constrained* traffic model.

In stochastic modeling, there are three common approaches: the first approach is to use stochastic processes such as Markov processes to model the traffic arrival process itself; the second one is a stochastic bounding approach, $e.g.$, the exponentially bounded burstiness (EBB) model [160] provides an upper bound on the probability of violating the LBAP constraint; the third approach uses large deviations theory, specifically, the asymptotic log-moment generating function of the traffic process, to characterize the traffic. Effective bandwidth and self-similar traffic model are two important stochastic models.

In this book, we focus on large-deviations-theoretical approach to model both the source and the channel. Hence, in the following, we introduce large-deviations-theoretical modeling, $i.e.$, theory of effective bandwidth.

2.2.1 Theory of Effective Bandwidth for Exponential Processes

The stochastic behavior of a source traffic process can be modeled asymptotically by its effective bandwidth, if the traffic is an exponential process [167].

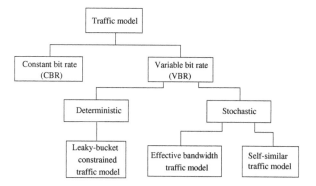

Figure 2: Classification of traffic models.

We will define exponential processes shortly. Next, we present the theory of effective bandwidth using Chang's framework [22, 23, 24, 167].

Consider an arrival process $\{A(t),\, t \geq 0\}$ where $A(t)$ represents the amount of source data (in bits) over the time interval $[0,\, t)$. Assume that sample paths of $A(t)$ are right continuous with left limits. For any $t \geq t_0 \geq 0$, let $A(t_0, t) = A(t) - A(t_0)$. Hence $A(t_0, t)$ denotes the cumulative arrivals over the time interval $[t_0, t)$. Assume that for $u \geq 0$ and $t \geq 0$,

$$\Lambda(u, t) = \frac{1}{t} \sup_{s \geq 0} \log E[e^{uA(s, s+t)}], \tag{2}$$

is bounded. Assume further that the asymptotic log-moment generating function of A, defined as

$$\Lambda(u) = \limsup_{t \to \infty} \Lambda(u, t) = \limsup_{t \to \infty} \frac{1}{t} \sup_{s \geq 0} \log E[e^{uA(s, s+t)}], \tag{3}$$

exists for all $u > 0$; a stochastic process satisfying this assumption is called an *exponential process*. Examples of exponential processes include all Markov processes, and all EBB processes. Then, the *effective bandwidth function* of an exponential process A is defined as

$$\alpha(u) = \frac{\Lambda(u)}{u}, \tag{4}$$

for all $u > 0$.

Under appropriate conditions, it can be shown [22] that $\alpha(u)$ is continuous and increasing in u, and that

$$\inf_{t \geq 0} \frac{1}{t} \sup_{s \geq 0} E[A(s, s+t)] = \alpha(0) \leq \alpha(u) \leq \alpha(\infty) = \inf_{t \geq 0} \frac{1}{t} \sup_{s \geq 0} ||A(s, s+t)||_\infty \quad (5)$$

where $||X||_\infty = \inf\{x : Pr(X > x) = 0\}$. Notice that $\alpha(0)$ can be regarded as the long term average rate of the arrival process A and $\alpha(\infty)$ the long term peak rate of A. Hence, $\alpha(u)$ is an increasing function from the long term average rate to the long term peak rate.

The effective capacity function $\alpha(u)$ is essential in characterizing the asymptotic behavior of a queueing system with an exponential arrival process A. Consider a queue of infinite buffer size served by a single server of rate r. Let $a(t)$ be the arrival rate to the queue at time t, and denote the queue length at time t by $Q(t)$. Using the theory of effective bandwidth, it can be shown that the probability of $Q(t)$ exceeding a threshold B satisfies [23]

$$\limsup_{B \to \infty} \frac{1}{B} \log \sup_{t \geq 0} Pr\{Q(t) \geq B\} \leq -u^* \quad (6)$$

where $u^* = \sup\{u > 0 : \alpha(u) < r\}$.

From (6), it is clear that for any $u > 0$, if $\alpha(u) < r$, then the asymptotic decay rate of the queue length process $Q(t)$, as defined by the left hand side of (6), is upper bounded by u. Informally speaking, the queue length decays exponentially at a rate of at least u. Owing to this fact, a process whose asymptotic log-moment generating function (3) exists for all $u > 0$, is referred to as an exponential process [167, page 24] to indicate that its queue length process has an exponentially decaying tail distribution.

The result in (6) can be made tight if stronger conditions on the arrival process are imposed. For example, if $A(s, t)$ is stationary, and

$$\Lambda(u) = \lim_{t \to \infty} \frac{1}{t} \log E[e^{uA(0,t)}], \quad (7)$$

exists and is differentiable for $u \in \mathbb{R}$, and $Q(0) = 0$, then one can show [24, page 291] that

$$\lim_{B \to \infty} \frac{1}{B} \log \sup_{t \geq 0} Pr\{Q(t) \geq B\} = \lim_{B \to \infty} \frac{1}{B} \log Pr\{Q(\infty) \geq B\} = -u^* \quad (8)$$

where $Q(\infty)$ denotes the steady state queue length, and u^* is the solution of the equation $\alpha(u) = r$. In other words, u^* can be obtained by

$$u^* = \alpha^{-1}(r) \quad (9)$$

where $\alpha^{-1}(\cdot)$ is the inverse function of $\alpha(u)$.

Under the aforementioned stronger conditions, the effective bandwidth functions of independent arrivals are additive [24, page 299]: let $\{a_i(t)\}$ ($1 \leq i \leq N$, $t \geq 0$) be N independent rate processes with effective bandwidths $\{\alpha_i(u)\}$, and let $a(t) = \sum_{i=1}^{N} a_i(t)$ ($t \geq 0$) be the aggregate rate process, then the effective bandwidth $\alpha(u)$ of the aggregate process is equal to $\sum_{i=1}^{N} \alpha_i(u)$, i.e., $\alpha(u) = \sum_{i=1}^{N} \alpha_i(u)$.

Eq. (8) indicates that the effective bandwidth $\alpha(u)$ of the arrival process characterizes the exact bandwidth requirement for achieving the event that $Pr\{Q(\infty) \geq q\} \leq e^{-uq}$ asymptotically, and suggests the following effective bandwidth approximation to the queue length tail distribution $Pr\{Q(\infty) \geq q\}$:

$$Pr\{Q(\infty) \geq q\} \approx e^{-u^*q} \tag{10}$$

In this section, we have introduced the basic results of effective bandwidth theory for exponential processes. We will use these results in later chapters.

2.3   Scheduling for Wireless Transmission

Packet scheduling is an important QoS provisioning mechanism at a network node. Compared with the scheduler design for the wired networks, the design of scheduling for wireless networks with QoS guarantees, is particularly challenging. This is because wireless channels have low reliability, and time varying signal strength, which may cause severe QoS violations. Further, the capacity of a wireless channel is severely limited, making efficient bandwidth utilization a priority.

In wired networks, the task of packet scheduling is to associate a packet with a time slot. In wireless networks, packet scheduling can be more general than that; its function is to schedule such resources as time slots, powers, data rates, channels, or combination of them, when packets are transmitted. (Note that a wired scheduler does not assign powers, data rates, and channels since packets are transmitted at a constant power, a constant data rate or link speed, and through one shared channel.)[1]   Specifically, based on the source characteristics, QoS requirements, channel states, and perhaps the queue lengths, a wireless scheduler assigns time slots, powers, data rates, channels, or combinations of them, to the packets for transmission. For example, in TDMA systems, time slots, powers, and data rates can be scheduled [30, 105];

---

[1]Here, we assume all flows share one wired channel/link. For the multiple shared channel case, a switch needs to be used.

in FDMA systems, channels (*i.e.*, frequencies) can be assigned [157] (see Figure 3(b)); in CDMA systems, powers, channels (*i.e.*, signature sequences), and data rates (*i.e.*, variable spreading factor) can be allotted [7, 43, 113, 137].

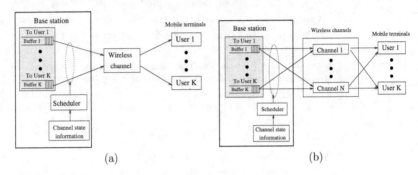

(a)   (b)

Figure 3: Wireless schedulers: (a) single channel, and (2) multiple channels.

A unique feature of wireless scheduling with QoS guarantees is its *channel state dependency*, *i.e.*, how to schedule the resources depends on the channel state (see Figure 3). This is necessary since without the knowledge about the channel state, it is impossible to guarantee QoS! A key difference between a wired scheduler and a wireless scheduler is that a wireless scheduler can utilize *asynchronous channel variations* or *multiuser diversity* [158] while a wired scheduler cannot.

Except the aforementioned differences, wireless and wired schedulers with QoS guarantees perform the same functions as below.

- *Isolation*: the scheduler supports the implementation of network service classes and provides isolation among these service classes in order to prevent one class from interfering with another class in achieving its QoS guarantees;

- *Sharing*: the scheduler controls bandwidth sharing among various service classes, and among flows in the same class, so that 1) statistical multiplexing gains can be exploited to efficiently utilize network resources, and 2) fair sharing of bandwidth among classes and sessions can be enforced.

There have been many proposals on scheduling with QoS constraints in wireless networks (see [46] for a survey). These schedulers fall into two classes: work-conserving and non-work-conserving. A work-conserving scheduler is never idle if there is a packet awaiting transmission. Examples include wireless fair queueing [87, 94, 106], modified largest weighted delay first (M-LWDF) [4], opportunistic transmission scheduling [83] and lazy packet scheduling [102]. In contrast, a non-work-conserving scheduler may be idle even if there is a backlogged packet in the queue because the packet may not be eligible for transmission. Examples are weighted round robin, the joint Knopp&Humblet/round robin (K&H/RR) scheduler [157] and the reference channel (RC) scheduler [158]. A non-work-conserving scheduler is an important component of a hierarchical scheduler supporting both QoS-assured flows and best-effort traffic. Such a hierarchical scheduler for wireless transmission was described in [152].

Next we briefly overview the pros and cons of some representative scheduling algorithms mentioned above.

Wireless fair queueing schemes [87, 94, 106] are aimed at applying wired fair queueing [99] to wireless networks. The objective of these schemes is to provide fairness, while providing loose QoS guarantees. Although these schedulers make decisions based on the channel state information (*i.e.*, good or bad channel), they do not exploit asynchronous channel variations to improve efficiency since packets destined to different users are transmitted at the same bit-rate.

The M-LWDF algorithm [4] and the opportunistic transmission scheduling [83] optimize a certain QoS parameter or utility index. They both exploit asynchronous channel variations and allow different user to transmit at different bit-rate or signal-to-interference-noise ratio (SINR), so that higher efficiency can be achieved. However, they do not provide the explicit QoS guarantees such as data rate, delay bound, and delay-bound violation probability.

The lazy packet scheduling [102] is targeted at minimizing energy, subject to a delay constraint. The scheme only considers AWGN channels and thus allows for a deterministic delay bound, unlike fading channels and the general statistical QoS considered in this book.

Time-division scheduling (or weighted round robin) has been proposed for 3-G WCDMA [60, page 226]. However, their proposal did not provide methods on how to use time-division scheduling to support statistical QoS guarantees explicitly.

To address the limitation of the above scheduling algorithms, *i.e.*, inability of provisioning explicit QoS, we present the K&H/RR scheduler in Chapter 5 and the reference channel scheduler in Chapter 6, respectively.

2.4   Call Admission Control in Wireless Networks

The objective of call admission control (CAC) is to provide QoS guarantees for individual connections while efficiently utilizing network resources. Specifically, a CAC algorithm makes the following decision:

> Given a call arriving to a network, can it be admitted by the network, with its requested QoS satisfied and without violating the QoS guarantees made to the existing connections?

The decision is made based on the availability of network resources as well as traffic specifications and QoS requirements of the users. If the decision is affirmative, necessary network resources need to be reserved to support the QoS. Hence, CAC is closely related to channel allocation, base station assignment, scheduling, power control, and bandwidth reservation. For example, whether the channel assignment is dynamic or fixed will result in different CAC algorithms.

The CAC problem can be formulated as an optimization problem, *i.e.*, maximize the network efficiency/utility/revenue subject to the QoS constraints of connections. The QoS constraints could be signal-to-interference ratio (SIR), the ratio of bit energy to interference density E_b/I_0, bit error rate (BER), call dropping probability, or connection-level QoS (such as a data rate, delay bound, and delay-bound violation probability triplet). For example, a CAC problem can be maximizing the number of users admitted or minimizing the blocking probability, subject to the BER violation probability not more than a required value ε_1, *i.e.*,

$$\text{maximize the number of users admitted} \qquad (11)$$

$$\text{or minimize the blocking probability} \qquad (12)$$

$$\text{subject to} \quad \Pr\{ \text{ BER} > \text{BER}_{th}\} \leq \varepsilon_1 \qquad (13)$$

where BER_{th} denotes a threshold for the BER. The constraint (13) can be replaced by

$$\text{subject to} \quad \text{the dropping probability} \leq \varepsilon_2 \qquad (14)$$

where the value of ε_2 may be different from that of ε_1.

CAC can also be used to provide priority to some service classes, or to enforce some policies like fair resource sharing, which includes complete sharing, complete partitioning, and threshold-based sharing.

There have been many algorithms on CAC in wireless networks (refer to [1] for a survey). These CAC algorithms may differ in admission criteria; they may be centralized or distributed; they may use global (all-cell) or local (single-cell) information about resource availability and interference levels to make admission decisions. The design of distributed CAC for cellular networks is not an easy task since intra-cell and inter-cell interference needs to be considered. The associated intra-cell and inter-cell resource allocation are complicated due to the interference.

A typical admission criterion is SIR. For example, Ref. [84] employs SIR to define a measure called *residual capacity*, and uses it as the admission criterion: if the residual capacity is positive, accept the new call; otherwise, reject it. Ref. [45] uses the concept of effective bandwidth[2] to measure whether the signal to interference density ratio (SIDR) can be satisfied for each class with certain probability. There, SIDR is defined as $r_s \times E_b/I_0$, where r_s denotes the source data rate in bits/sec. If the total effective bandwidth including that for the new call, is less than the available bandwidth, the new call will be accepted; otherwise, it will be rejected.

Another important admission criterion is transmitted or received power. In [73], a new call is admitted if the total transmitted power does not exceed a preset value. In [74], the CAC uses the 95-percentile of the total received power as the admission criterion.

However, none of the existing CAC algorithms provides explicit connection-level QoS guarantees such as a data rate, delay bound, and delay-bound violation probability triplet. In this book, we propose CAC algorithms that are capable of providing connection-level QoS guarantees explicitly. Such CAC schemes will be presented in Chapters 5 and 6.

2.5   Wireless Channel Modeling

Figure 4 shows a wireless communication system. The data source generates packets and the packets are first put into a buffer to accommodate the mismatch between the source rate and the time-varying wireless channel capacity. Then the packets traverse a channel encoder, a modulator, a wireless channel, a demodulator, a channel decoder, a network access device, and finally reach the data sink.

As shown in Figure 4, one can model the communication channel at different layers as below

- Radio-layer channel: is the part between the output of the modulator and the input of the demodulator.

---

[2]Note that the effective bandwidth defined in [45] is different from that defined in Section 2.2.1.

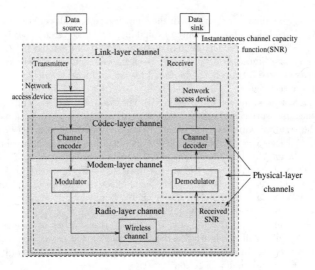

Figure 4: A wireless communication system and associated channel models.

- Modem-layer channel: is the part between the output of the channel encoder and the input of the channel decoder.

- Codec-layer channel: is the part between the output of the network access device at the transmitter, and the input of the network access device at the receiver.

- Link-layer channel: is the part between the output of the data source and the input of the data sink.

The above radio-layer, the modem-layer, and the codec-layer channels can all be regarded as physical-layer channels.

As shown in Figure 5, radio-layer channel models can be classified into two categories: large-scale path loss and small-scale fading. Large-scale path loss models, also called propagation models, characterize the underlying physical mechanisms (*i.e.*, reflection, diffraction, scattering) for specific paths. These models specify signal attenuation as a function of distance, which is affected by prominent terrain contours (buildings, hills, forests, etc.) between the transmitter and the receiver. Path loss models describe the mean signal attenuation vs. distance in a deterministic fashion (*e.g.*, nth-power law [109]), and also the statistical variation about the mean (*e.g.*, log-normal distribution [109]).

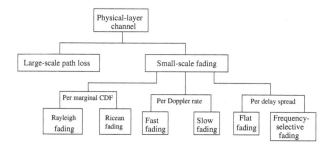

Figure 5: Classification of physical-layer channel models.

Small-scale fading models describe the characteristics of generic radio paths in a statistical fashion. Small-scale fading refers to the dramatic changes in signal amplitude and phase that can be experienced as a result of small changes (as small as a half-wavelength) in the spatial separation between the receiver and the transmitter [119]. Small-scale fading can be slow or fast, depending on the Doppler rate. Small-scale fading can also be flat or frequency-selective, depending on the delay spread of the channel. The statistical time-varying nature of the envelope of a flat-fading signal is characterized by distributions such as Rayleigh, Ricean, Nakagami, etc. [109]. Uncorrelated scattering is often assumed, to extend these distributions to the frequency-selective case. The large-scale path loss and small-scale fading together characterize the received signal power over a wide range of distances.

A modem-layer channel can be modeled by a finite-state Markov chain [165], whose states are characterized by different bit error rates (BER). For example, in [165], a Rayleigh fading with certain Doppler spectrum is converted to a BER process, modeled by a finite-state Markov chain. The idea is the following: 1) quantize the continuous Rayleigh random variable into a discrete random variable, based on certain optimal criterion (*e.g.*, minimum mean squared error), 2) map the resulting discrete random variable or SNR to discrete BER, for a given modulation scheme (say, binary phase shift keying), and 3) estimate the state transition probabilities, which reflect the Doppler spectrum. This procedure gives the states (*i.e.*, BER's) and the transition probability matrix of the Markov chain.

A codec-layer channel can also be modeled by a finite-state Markov chain, whose states can be characterized by different data-rates [71], or symbol being error-free/in-error, or channel being good/bad [170]. The two state Markov chain model with good/bad states [170] is widely used in analyzing the performance of upper layer protocols such as TCP [171]. If the decoder uses hard decisions from the demodulator/detector, a codec-layer channel model can be easily obtained from a modem-layer channel model. For example, the good/bad channel model can be derived from a finite-state Markov chain with BER's as the states in the following way: first compute symbol error probability from BER; then decide the channel being good if the symbol error probability is less than a preset threshold, otherwise decide the channel being bad. The resulting good/bad channel process is a two state Markov chain.

Radio-layer channel models provide a quick estimate of the performance of wireless communications systems (*e.g.*, symbol error rate vs. signal-to-noise ratio (SNR)). However, radio-layer channel models cannot be easily translated into complex QoS guarantees for a connection, such as bounds on delay violation probability and packet loss ratio. The reason is that, these complex QoS requirements need an analysis of the queueing behavior of the connection, which is hard to extract from radio-layer models [156]. Thus it is hard to use radio-layer models in QoS support mechanisms, such as admission control and resource reservation.

Finite-state Markov chain models for a modem-layer or codec-layer channel also require a queueing analysis of very high complexity to obtain connection-level QoS such as a data rate, delay bound, and delay-bound violation probability triplet. We will show this through an example in Section 5.2.3.

Recognizing that the limitation of the physical-layer channel models in QoS support, is the difficulty in analyzing queues, we propose moving the channel model up the protocol stack, from the physical-layer to the link-layer. The resulting link-layer channel model is called *effective capacity* model and will be presented in Chapter 4.

CHAPTER 3

END-SYSTEM-BASED QOS SUPPORT: AN OVERVIEW

In this chapter, we overview the issues and challenges in end-system-based QoS support, and present a survey of related work. Since end-system-based QoS support is typically application specific, for illustration purpose, we choose real-time video transport over the Internet as the target application. The end-system-based techniques surveyed here are general and hence also applicable to wireless video transport.

3.1 Introduction

Unicast and multicast delivery of real-time video are important building blocks of many Internet multimedia applications, such as Internet television, video conferencing, distance learning, digital libraries, tele-presence, and video-on-demand. Transmission of real-time video has bandwidth, delay and loss requirements. However, there is no QoS guarantee for video transmission over the current Internet. In addition, for video multicast, the heterogeneity of the networks and receivers makes it difficult to achieve bandwidth efficiency and service flexibility. Therefore, there are many challenging issues that need to be addressed in designing protocols and mechanisms for Internet video transmission.

We list the challenging QoS issues as follows.

1. **Bandwidth.** To achieve acceptable presentation quality, transmission of real-time video typically has minimum bandwidth requirement (say, 28 kb/s). However, the current Internet does not provide bandwidth reservation to meet such a requirement. Furthermore, since traditional routers typically do not actively participate in congestion control [21], excessive traffic can cause congestion collapse, which can further degrade the throughput of real-time video.

2. **Delay.** In contrast to data transmission, which are usually not subject to strict delay constraints, real-time video requires bounded end-to-end delay (say, 1 second). That is, every video packet must arrive at the

39

destination in time to be decoded and displayed. This is because real-time video must be played out continuously. If the video packet does not arrive timely, the playout process will pause, which is annoying to human eyes. In other words, the video packet that arrives beyond a time constraint is useless and can be considered lost. Although real-time video requires timely delivery, the current Internet does not offer such a delay guarantee. In particular, the congestion in the Internet could incur excessive delay, which exceeds the delay requirement of real-time video.

3. **Loss.** Loss of packets can potentially make the presentation displeasing to human eyes, or, in some cases, make the presentation impossible. Thus, video applications typically impose some packet loss requirements. Specifically, the packet loss ratio is required to be kept below a threshold (say, 1%) to achieve acceptable visual quality. Although real-time video has a loss requirement, the current Internet does not provide any loss guarantee. In particular, the packet loss ratio could be very high during network congestion, causing severe degradation of video quality.

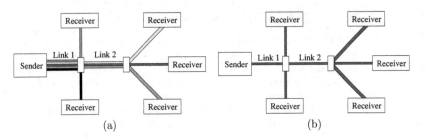

(a)   (b)

Figure 6: (a) Unicast video distribution using multiple point-to-point connections. (b) Multicast video distribution using point-to-multipoint transmission.

Besides the above QoS problems, for video multicast applications, there is another challenge coming from the *heterogeneity* problem. Before addressing the heterogeneity problem, we first describe the advantages and disadvantages of unicast and multicast. The unicast delivery of real-time video uses point-to-point transmission, where only one sender and one receiver are involved. In contrast, the multicast delivery of real-time video uses point-to-multipoint transmission, where one sender and multiple receivers are involved. For applications such as video conferencing and Internet television, delivery using multicast can achieve high bandwidth efficiency since the receivers can share links. On the other hand, unicast delivery of such applications is inefficient in terms of bandwidth utilization. An example is give in Fig. 6, where, for unicast, five copies of the video content flow across Link 1 and three copies flow across Link 2 as shown in Fig. 6(a). In contrast, the multicast removes this replication. That is, there is only one copy of the video content traversing any link in the network (Fig. 6(b)), resulting in substantial bandwidth savings. However, the efficiency of multicast is achieved at the cost of losing the service flexibility of unicast (*i.e.*, in unicast, each receiver can individually negotiate service parameters with the source). Such lack of flexibility in multicast can be problematic in a heterogeneous network environment, which we elaborate as follows.

Heterogeneity. There are two kinds of heterogeneity, namely, *network heterogeneity* and *receiver heterogeneity*. Network heterogeneity refers to the subnetworks in the Internet having unevenly distributed resources (*e.g.*, processing, bandwidth, storage and congestion control policies). Network heterogeneity could make different user experience different packet loss/delay characteristics. Receiver heterogeneity means that receivers have different or even varying latency requirements, visual quality requirement, and/or processing capability. For example, in live multicast of a lecture, participants who want to ask questions and interact with the lecturer desire stringent real-time constraints on the video while passive listeners may be willing to sacrifice latency for higher video quality.

The sharing nature of multicast and the heterogeneity of networks and receivers sometimes present a conflicting dilemma. For example, the receivers in Fig. 6(b) may attempt to request for different video quality with different bandwidth. But only one copy of the video content is sent out from the source. As a result, all the receivers have to receive the same video content with the same quality. It is thus a challenge to design a multicast mechanism that not only achieves efficiency in network bandwidth but also meets the various requirements of the receivers.

To address the above technical issues, extensive research based on the end system-based approach has been conducted and various solutions have been proposed. This chapter is aimed at giving the reader a big picture of this challenging area and identifying a design space that can be explored by video application designers. We take a holistic approach to present solutions from both transport and compression perspectives. By transport perspective, we refer to the use of control/processing techniques without regard of the specific video semantics. In other words, these control/processing techniques are applicable to generic data. By compression perspective, we mean employing signal processing techniques with consideration of the video semantics on the compression layer. With the holistic approach, we next present a framework, which consists of two components, namely, *congestion control* and *error control*.

1. **Congestion control.** Bursty loss and excessive delay have devastating effect on video presentation quality and they are usually caused by network congestion. Thus, congestion control is required to reduce packet loss and delay. One congestion control mechanism is *rate control* [17]. Rate control attempts to minimize network congestion and the amount of packet loss by matching the rate of the video stream to the available network bandwidth. In contrast, without rate control, the traffic exceeding the available bandwidth would be discarded in the network. To force the source to send the video stream at the rate dictated by the rate control algorithm, rate adaptive video encoding [150] or rate shaping [44] is required.

2. **Error control.** The purpose of congestion control is to prevent packet loss. However, packet loss is unavoidable in the Internet and may have significant impact on perceptual quality. Thus, other mechanisms must be in place to maximize video presentation quality in presence of packet loss. Such mechanisms include error control mechanisms, which can be classified into four types, namely, *forward error correction (FEC), retransmission, error-resilience,* and *error concealment*.

The remainder of this chapter is organized as follows. Section 3.2 presents the approaches for congestion control. In Section 3.3, we describe the mechanisms for error control. Section 3.4 summarizes this chapter.

3.2   Congestion Control

There are three mechanisms for congestion control: rate control, rate adaptive video encoding, and rate shaping. Rate control follows the transport approach; rate adaptive video encoding follows the compression approach; rate shaping could follow either the transport approach or the compression approach.

For the purpose of illustration, we present an architecture including the three congestion control mechanisms in Fig. 7, where the rate control is a source-based one (*i.e.*, the source is responsible for adapting the rate). Although the architecture in Fig. 7 is targeted at transporting live video, this architecture is also applicable to stored video if the rate adaptive encoding is excluded. At the sender side, the compression layer compresses the live video based on a rate adaptive encoding algorithm. After this stage, the compressed video bit-stream is first filtered by the rate shaper and then passed through

42

the RTP/UDP/IP layers before entering the Internet, where RTP is Real-time Transport Protocol [114]. Packets may be dropped inside the Internet (due to congestion) or at the destination (due to excess delay). For packets that are successfully delivered to the destination, they first pass through the IP/UDP/RTP layers before being decoded at the video decoder.

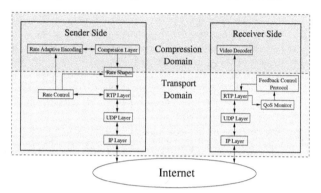

Figure 7: A layered architecture for transporting real-time video.

Under our architecture, a QoS monitor is maintained at the receiver side to infer network congestion status based on the behavior of the arriving packets, *e.g.*, packet loss and delay. Such information is used in the feedback control protocol, which sends information back to the video source. Based on such feedback information, the rate control module estimates the available network bandwidth and conveys the estimated network bandwidth to the rate adaptive encoder or the rate shaper. Then, the rate adaptive encoder or the rate shaper regulates the output rate of the video stream according to the estimated network bandwidth. It is clear that the source-based congestion control must include: (1) rate control, and (2) rate adaptive video encoding or rate shaping.

We organize the rest of this section as follows. In Section 3.2.1, we survey the approaches for rate control. Section 3.2.2 describes basic methods for rate-adaptive video encoding. In Section 3.2.3, we classify methodologies for rate shaping and summarize representative schemes.

3.2.1 Rate Control: A Transport Approach

Since TCP retransmission introduces delays that may not be acceptable for real-time video applications, UDP is usually employed as the transport protocol for real-time video streams [150]. However, UDP is not able to provide congestion control and overcome the lack of service guarantees in the Internet. Therefore, it is necessary to implement a control mechanism on the upper layer (higher than UDP) to prevent congestion.

There are two types of control for congestion prevention: one is *window-based* [67] and the other is *rate-based* [129]. The window-based control such as TCP works as follows: it probes for the available network bandwidth by slowly increasing a congestion window (used to control how much data is outstanding in the network); when congestion is detected (indicated by the loss of one or more packets), the protocol reduces the congestion window greatly. The rapid reduction of the window size in response to congestion is essential to avoid network collapse. On the other hand, the rate-based control sets the sending rate based on the estimated available bandwidth in the network; if the estimation of the available network bandwidth is relatively accurate, the rate-based control could also prevent network collapse.

Since the window-based control like TCP typically couples retransmission which can introduce intolerable delays, the rate-based control (*i.e.*, rate control) is usually employed for transporting real-time video [150]. Existing rate control schemes for real-time video can be classified into three categories, namely, source-based, receiver-based, and hybrid rate control, which are described in Sections 3.2.1.1 to 3.2.1.3, respectively.

3.2.1.1   Source-based Rate Control

Under the source-based rate control, the sender is responsible for adapting the transmission rate of the video stream. The source-based rate control can minimize the amount of packet loss by matching the rate of the video stream to the available network bandwidth. In contrast, without rate control, the traffic exceeding the available bandwidth could be discarded in the network.

Typically, feedback is employed by source-based rate control mechanisms to convey the changing status of the Internet. Based upon the feedback information about the network, the sender could regulate the rate of the video stream. The source-based rate control can be applied to both unicast [150] and multicast [15].

For unicast video, the existing source-based rate control mechanisms can be classified into two approaches, namely, the probe-based approach and the model-based approach, which are presented as follows.

Probe-based approach. Such an approach is based on probing experiments. Specifically, the source probes for the available network bandwidth by adjusting the sending rate so that some QoS requirements are met, *e.g.*, packet loss ratio p is below a certain threshold P_{th} [150]. The value of P_{th} is determined according to the minimum video perceptual quality required by the receiver. There are two ways to adjust the sending rate: Additive Increase and

44

Multiplicative Decrease (AIMD) [150], and Multiplicative Increase and Multiplicative Decrease (MIMD) [129]. The probe-based rate control could avoid congestion since it always tries to adapt to the congestion status, *e.g.*, keep the packet loss at an acceptable level.

For the purpose of illustration, we briefly describe the source-based rate control based on additive increase and multiplicative decrease. The AIMD rate control algorithm is shown as follows [150].

if $(p \le P_{th})$

$$r := \min\{(r + \text{AIR}), \text{MaxR}\};$$

else

$$r := \max\{(\alpha \times r), \text{MinR}\}.$$

where p is the packet loss ratio; P_{th} is the threshold for the packet loss ratio; r is the sending rate at the source; AIR is the additive increase rate; MaxR and MinR are the maximum rate and the minimum rate of the sender, respectively; and α is the multiplicative decrease factor.

Packet loss ratio p is measured by the receiver and conveyed back to the sender. An example source rate behavior under the AIMD rate control is illustrated in Fig. 8.

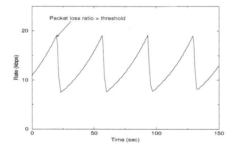

Figure 8: Source rate behavior under the AIMD rate control.

Model-based approach. Different from the probe-based approach, which implicitly estimates the available network bandwidth, the model-based approach attempts to estimate the available network bandwidth explicitly. This can be achieved by using a throughput model of a TCP connection, which is

45

characterized by the following formula [47]:

$$\lambda = \frac{1.22 \times MTU}{RTT \times \sqrt{p}},$$ (15)

where λ is the throughput of a TCP connection, MTU (Maximum Transit Unit) is the maximum packet size used by the connection, RTT is the round trip time for the connection, p is the packet loss ratio experienced by the connection. Under the model-based rate control, Eq. (15) can be used to determine the sending rate of the video stream. That is, the rate-controlled video flow gets its bandwidth share like a TCP connection. As a result, the rate-controlled video flow could avoid congestion in a way similar to that of TCP, and can co-exist with TCP flows in a "friendly" manner. Hence, the model-based rate control is also called "TCP friendly" rate control [139]. In contrast to this TCP friendliness, a flow without rate control can get much more bandwidth than a TCP flow when the network is congested. This may lead to possible starvation of competing TCP flows due to the rapid reduction of the TCP window size in response to congestion.

To compute the sending rate λ in Eq. (15), it is necessary for the source to obtain the MTU, RTT, and packet loss ratio p. The MTU can be found through the mechanism proposed by Mogul and Deering [92]. In the case when the MTU information is not available, the default MTU, $i.e.$, 576 bytes, will be used. The parameter RTT can be obtained through feedback of timing information. In addition, the receiver can periodically send the parameter p to the source in the time scale of the round trip time. Upon the receipt of the parameter p, the source estimates the sending rate λ and then a rate control action may be taken.

Single-channel multicast vs. unicast. For multicast under the source-based rate control, the sender uses a single channel or one IP multicast group to transport the video stream to the receivers. Thus, such multicast is called *single-channel multicast.*

For single-channel multicast, only the probe-based rate control can be employed [15]. A representative work is the IVS (INRIA Video-conference System) [15]. The rate control in IVS is based on additive increase and multiplicative decrease, which is summarized as follows. Each receiver estimates its packet loss ratio, based on which, each receiver can determine the network status to be in one of the three states: UNLOADED, LOADED, and CONGESTED. The source solicits the network status information from the receivers through probabilistic polling, which helps to avoid feedback implosion.[1]   In this way, the fraction of UNLOADED and CONGESTED receivers

---

[1]Feedback implosion means that there are too many feedback messages for the source to

46

can be estimated. Then, the source adjusts the sending rate according to the following algorithm.

if $(F_{con} > T_{con})$

$\quad r := \max(r/2, \text{MinR});$

else if $(F_{un} == 100\%)$

$\quad r := \min\{(r + \text{AIR}), \text{MaxR}\};$

where F_{con}, F_{un}, and T_{con} are fraction of CONGESTED receivers, fraction of UNLOADED receivers, and a preset threshold, respectively; r, MaxR, MinR, and AIR are the sending rate, the maximum rate, the minimum rate, and additive increase rate, respectively.

Single-channel multicast has good bandwidth efficiency since all the receivers share one channel (*e.g.*, the IP multicast group in Fig. 6(b)). But single-channel multicast is unable to provide service flexibility and differentiation to different receivers with diverse access link capacities, processing capabilities and interests.

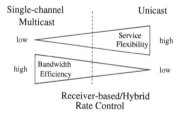

Figure 9: Trade-off between bandwidth efficiency and service flexibility.

On the other hand, multicast video, delivered through individual unicast streams (see Fig. 6(a)), can offer differentiated services to receivers since each receiver can individually negotiate the service parameters with the source. But the problem with unicast-based multicast video is bandwidth inefficiency.

Single-channel multicast and unicast-based multicast are two extreme cases shown in Fig. 9. To achieve good trade-off between bandwidth efficiency and service flexibility for multicast video, two mechanisms, namely, receiver-based and hybrid rate control, have been proposed, which we discuss as follows.

---

handle.

3.2.1.2 Receiver-based Rate Control

Under the receiver-based rate control, the receivers regulate the receiving rate of video streams by adding/dropping channels. In contrast to the sender-based rate control, the sender does not participate in rate control here. Typically, the receiver-based rate control is applied to layered multicast video rather than unicast video. This is primarily because the source-based rate control works reasonably well for unicast video and the receiver-based rate control is targeted at solving heterogeneity problem in the multicast case.

Before we address the receiver-based rate control, we first briefly describe layered multicast video as follows. At the sender side, a raw video sequence is compressed into multiple layers: a base layer (*i.e.*, Layer 0) and one or more enhancement layers (*e.g.*, Layers 1 and 2 in Fig. 10). The base layer can be independently decoded and it provides basic video quality; the enhancement layers can only be decoded together with the base layer and they further refine the quality of the base layer. This is illustrated in Fig. 10. The base layer consumes the least bandwidth (*e.g.*, 64 kb/s in Fig. 10); the higher the layer is, the more bandwidth the layer consumes (see Fig. 10). After compression, each video layer is sent to a separate IP multicast group. At the receiver side, each receiver subscribes to a certain set of video layers by joining the corresponding IP multicast group. In addition, each receiver tries to achieve the highest subscription level of video layers without incurring congestion. In the example shown in Fig. 11, each layer has a separate IP multicast group. Receiver 1 joins all three IP multicast groups. As a result, it consumes 1 Mb/s and receives all the three layers. Receiver 2 joins the two IP multicast groups for Layer 0 and Layer 1 with bandwidth usage of 256 kb/s. Receiver 3 only joins the IP multicast group for Layer 0 with bandwidth consumption of 64 kb/s.

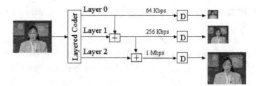

Figure 10: Layered video encoding/decoding. D denotes the decoder.

Like the source-based rate control, we classify existing receiver-based rate control mechanisms into two approaches, namely, the probe-based approach and the model-based approach, which are presented as follows.

Probe-based approach. This approach was first employed in Receiver driven Layered Multicast (RLM) [91]. Basically, the probe-based rate control consists

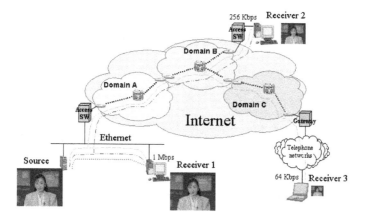

Figure 11: IP multicast for layered video.

of two parts:

1. When no congestion is detected, a receiver probes for the available bandwidth by joining a layer, which leads to an increase of its receiving rate. If no congestion is detected after the joining, the join-experiment is considered "successful". Otherwise, the receiver drops the newly added layer.

2. When congestion is detected, the receiver drops a layer, resulting in reduction of its receiving rate.

The above control has a potential problem when the number of receivers becomes large. If each receiver carries out the above join-experiment independently, the aggregate frequency of such experiments increases with the number of receivers. Since a failed join-experiment could incur congestion to the network, an increase of join-experiments could aggravate network congestion.

To minimize the frequency of join-experiments, a shared learning algorithm was proposed in [91]. The essence of the shared learning algorithm is to have a receiver multicast its intent to the group before it starts a join-experiment. In this way, each receiver can learn from other receivers' failed join-experiments, resulting in a decrease of the number of failed join-experiments.

The shared learning algorithm in [91] requires each receiver to maintain a comprehensive group knowledge base, which contains the results of all the join-experiments for the multicast group. In addition, the use of multicasting to update the comprehensive group knowledge base may decrease usable bandwidth on low-speed links and lead to lower quality for receivers on these

49

links. To reduce message processing overhead at each receiver and to decrease bandwidth usage of the shared learning algorithm, a hierarchical rate control mechanism called Layered Video Multicast with Retransmissions (LVMR) [80] was proposed. The methodology of the hierarchical rate control is to partition the comprehensive group knowledge base, organize the partitions in a hierarchical way and distribute relevant information (rather than all the information) to the receivers. In addition, the partitioning of the comprehensive group knowledge base allows multiple experiments to be conducted simultaneously, making it faster for the rate to converge to the stable state. Although the hierarchical rate control could reduce control protocol traffic, it requires installing agents in the network so that the comprehensive group knowledge base can be partitioned and organized in a hierarchical way.

Model-based approach. Unlike the probe-based approach which implicitly estimates the available network bandwidth through probing experiments, the model-based approach attempts to explicitly estimate the available network bandwidth. The model-based approach is based on the throughput model of a TCP connection, which was described in Section 3.2.1.1.

Fig. 12 shows the flow chart of the basic model-based rate control executed by each receiver, where γ_i is the transmission rate of Layer i. In the algorithm, it is assumed that each receiver knows the transmission rate of all the layers. For the ease of description, we divide the algorithm into the following steps.

Initialization: A receiver starts with subscribing the base layer (*i.e.*, Layer 0) and initializes the variable L to 0. The variable L represents the highest layer currently subscribed.

Step 1: Receiver estimates MTU, RTT, and packet loss ratio p for a given period. The MTU can be found through the mechanism proposed by Mogul and Deering [92]. Packet loss ratio p can be easily obtained. However, the RTT cannot be measured through a simple feedback mechanism due to feedback implosion problem. A mechanism [130], based on RTCP protocol, has been proposed to estimate the RTT.

Step 2: Upon obtaining MTU, RTT, and p for a given period, the target rate λ can be computed through Eq. (15).

Step 3: Upon obtaining λ, a rate control action can be taken. If $\lambda < \gamma_0$, drop the base layer and stop receiving video (the network cannot deliver even the base layer due to congestion); otherwise, determine L', the largest integer such that $\sum_{i=0}^{L'} \gamma_i \leq \lambda$. If $L' > L$, add the layers from Layer $L+1$ to Layer L', and Layer L' becomes the highest layer currently subscribed (let $L = L'$); if $L' < L$, drop layers from Layer $L' + 1$ to Layer L, and

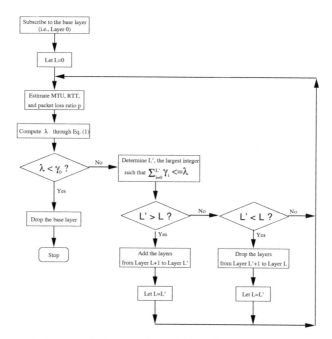

Figure 12: Flow chart of the basic model-based rate control for a receiver.

Layer L' becomes the highest layer currently subscribed (let $L = L'$).
Return to Step 1.

The above algorithm has a potential problem when the number of receivers
becomes large. If each receiver carries out the rate control independently, the
aggregate frequency of join-experiments increases with the number of receivers.
Since a failed join-experiment could incur congestion to the network, an increase of join-experiments could aggravate network congestion. To coordinate
the joining/leaving actions of the receivers, a scheme based on synchronization points [134] was proposed. With small protocol overhead, the proposed
scheme in [134] helps to reduce the frequency and duration of join-experiments,
resulting in a smaller possibility of congestion.

3.2.1.3 Hybrid Rate Control

Under the hybrid rate control, the receivers regulate the receiving rate of video
streams by adding/dropping channels while the sender also adjusts the transmission rate of each channel based on feedback information from the receivers.
Since the hybrid rate control consists of rate control at both the sender and a
receiver, previous approaches described in Sections 3.2.1.1 and 3.2.1.2 can be
employed.

The hybrid rate control is targeted at multicast video and is applicable to
both layered video [120] and non-layered video [25]. Different from the source-
based rate control framework where the sender uses a single channel, the hybrid
rate control framework uses multiple channels. On the other hand, different
from the receiver-based rate control framework where the rate for each channel
is constant, the hybrid rate control enables the sender to dynamically change
the rate for each channel based on congestion status.

One representative work using hybrid rate control is destination set grouping (DSG) protocol [25]. Before we present the DSG protocol, we first briefly
describe the architecture associated with DSG. At the sender side, a raw
video sequence is compressed into multiple streams (called replicated adaptive streams), which carry the same video information with different rate and
quality. Different from layered video, each stream in DSG can be decoded
independently. After compression, each video stream is sent to a separate IP
multicast group. At the receiver side, each receiver can choose a multicast
group to join by taking into account of its capability and congestion status.
The receivers also send feedback to the source, and the source uses this feedback to adjust the transmission rate for each stream.

The DSG protocol consists of two main components:

1. **Rate control at the source.** For each stream, the rate control at the

source is essentially the same as that used in IVS (see Section 3.2.1.1). But the feedback control for each stream works independently.

2. **Rate control at a receiver.** A receiver can change its subscription and join a higher or lower quality stream based on network status, *i.e.*, the fraction of UNLOADED, LOADED and CONGESTED receivers. The mechanism to obtain the fraction of UNLOADED, LOADED and CONGESTED receivers is similar to that used in IVS. The rate control at a receiver takes the probe-based approach as presented in Section 3.2.1.2.

3.2.2 Rate-adaptive Video Encoding: A Compression Approach

Rate-adaptive video encoding has been studied extensively for various standards and applications, such as video conferencing with H.261 and H.263 [89, 144], storage media with MPEG-1 and MPEG-2 [39, 77, 125], real-time transmission with MPEG-1 and MPEG-2 [40, 62], and the recent object-based coding with MPEG-4 [133, 150]. The objective of a rate-adaptive encoding algorithm is to maximize the perceptual quality under a given encoding rate.[2]

Such adaptive encoding can be achieved by the alteration of the encoder's quantization parameter (QP) and/or the alteration of the video frame rate.

Traditional video encoders (*e.g.*, H.261, MPEG-1/2) typically rely on altering the QP of the encoder to achieve rate adaptation. These encoding schemes must perform coding with constant frame rates. This is because even a slight reduction in frame rate can substantially degrade the perceptual quality at the receiver, especially during a dynamic scene change. Since altering the QP is not enough to achieve very low bit-rate, these encoding schemes may not be suitable for very low bit-rate video applications.

On the contrary, MPEG-4 and H.263 coding schemes are suitable for very low bit-rate video applications since they allow the alteration of the frame rate. In fact, the alteration of the frame rate is achieved by frame-skip.[3] Specifically, if the encoder buffer is in danger of overflow (*i.e.*, the bit budget is over-used by the previous frame), a complete frame can be skipped at the encoder. This will allow the coded bits of the previous frames to be transmitted during the time period of this frame, therefore reducing the buffer level (*i.e.*, keeping the encoded bits within the budget).

In addition, MPEG-4 is the first international standard addressing the coding of video objects (VO's) (see Fig. 13) [64]. With the flexibility and efficiency provided by coding video objects, MPEG-4 is capable of addressing

---

[2]The given encoding rate can be either fixed or dynamically changing based on the network congestion status.

[3]Skipping a frame means that the frame is not encoded.

interactive content-based video services as well as conventional stored and live video [108]. In MPEG-4, a frame of a video object is called a video object plane (VOP), which is encoded separately. Such isolation of video objects provides us with much greater flexibility to perform adaptive encoding. In particular, we can dynamically adjust target bit-rate distribution among video objects, in addition to the alteration of QP on each VOP (such a scheme is proposed in [150]). This can upgrade the perceptual quality for the regions of interest (*e.g.*, head and shoulder) while lowering the quality for other regions (*e.g.*, background).

Figure 13: An example of video object (VO) concept in MPEG-4 video. A video frame (left) is segmented into two VO planes where VO1 (middle) is the background and VO2 (right) is the foreground.

For all the video coding algorithms, a fundamental problem is how to determine a suitable QP to achieve the target bit-rate. The rate-distortion (R-D) theory is a powerful tool to solve this problem. Under the R-D framework, there are two approaches for encoding rate control in the literature: the model-based approach and the operational R-D based approach. The model-based approach assumes various input distributions and quantizer characteristics [26, 150]. Under this approach, closed-form solutions can be obtained by using continuous optimization theory. On the other hand, the operational R-D based approach considers practical coding environments where only a finite set of quantizers is admissible [62, 77, 125, 144]. Under the operational R-D based approach, the admissible quantizers are used by the rate control algorithm to determine the optimal strategy to minimize the distortion under the constraint of a given bit budget. The optimal discrete solutions can be found through applying integer programming theory.

3.2.3 Rate Shaping

Rate shaping is a technique to adapt the rate of compressed video bit-streams to the target rate constraint. A rate shaper is an interface (or filter) between the encoder and the network, with which the encoder's output rate can be matched to the available network bandwidth. Since rate shaping does not require interaction with the encoder, rate shaping is applicable to any video coding scheme and is applicable to both live and stored video. Rate shaping can be achieved through two approaches: one is from the transport perspective [59, 121, 168] and the other is from the compression perspective [44].

A representative mechanism from the transport perspective is *server selective frame discard* [168]. The server selective frame discard is motivated by the following fact. Usually, a server transmits each frame without any awareness of the available network bandwidth and the client buffer size. As a result, the network may drop packets if the available bandwidth is less than required, which leads to frame losses. In addition, the client may also drop packets that arrive too late for playback. This causes wastage of network bandwidth and client buffer resources. To address this problem, the selective frame discard scheme preemptively drops frames at the server in an intelligent manner by considering available network bandwidth and client QoS requirements. The selective frame discard has two major advantages. First, by taking the network bandwidth and client buffer constraints into account, the server can make the best use of network resources by selectively discarding frames in order to minimize the likelihood of future frames being discarded, thereby increasing the overall quality of the video delivered. Second, unlike frame dropping in the network or at the client, the server can also take advantage of application-specific information such as regions of interest and group of pictures (GOP) structure, in its decision in discarding frames. As a result, the server optimizes the perceived quality at the client while maintaining efficient utilization of the network resources.

A representative mechanism from the compression perspective is dynamic rate shaping [44]. Based on the R-D theory, the dynamic rate shaper selectively discards the Discrete Cosine Transform (DCT) coefficients of the high frequencies so that the target rate can be achieved. Since human eyes are less sensitive to higher frequencies, the dynamic rate shaper selects the highest frequencies and discards the DCT coefficients of these frequencies until the target rate is met.

Congestion control attempts to prevent packet loss by matching the rate of video streams to the available bandwidth in the network. However, packet loss is unavoidable in the Internet and may have significant impact on perceptual quality. Therefore, we need other mechanisms to maximize the video presentation quality in presence of packet loss. Such mechanisms include error control mechanisms, which are presented in the next section.

3.3 Error Control

In the Internet, packets may be dropped due to congestion at routers, they may be mis-routed, or they may reach the destination with such a long delay as to be considered useless or lost. Packet loss may severely degrade the visual presentation quality. To enhance the video quality in presence of packet loss, error control mechanisms have been proposed.

For certain types of data (such as text), packet loss is intolerable while delay is acceptable. When a packet is lost, there are two ways to recover the packet: the corrupted data must be corrected by traditional FEC (*i.e.*, channel coding), or the packet must be retransmitted. On the other hand, for real-time video, some visual quality degradation is often acceptable while delay must be bounded. This feature of real-time video introduces many new error control mechanisms, which are applicable to video applications but not applicable to traditional data such as text. Basically, the error control mechanisms for video applications can be classified into four types, namely, FEC, retransmission, error-resilience, and error concealment. FEC, retransmission, and error-resilience are performed at both the source and the receiver side, while error concealment is carried out only at the receiver side. Fig. 14 shows the location of each error control mechanism in a layered architecture. As shown in Fig. 14, retransmission recovers packet loss from the transport perspective; error-resilience and error concealment deal with packet loss from the compression perspective; and FEC falls in both transport and compression domains. For the rest of this section, we present FEC, retransmission, error-resilience, and error concealment, respectively.

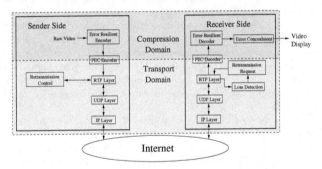

Figure 14: An architecture for error control mechanisms.

56

3.3.1 FEC

The use of FEC is primarily because of its advantage of small transmission delay, compared with TCP [35]. The principle of FEC is to add extra (redundant) information to a compressed video bit-stream so that the original video can be reconstructed in presence of packet loss. Based on the kind of redundant information to be added, the existing FEC schemes can be classified into three categories: (1) channel coding, (2) source coding-based FEC, and (3) joint source/channel coding, which will be presented in Sections 3.3.1.1 to 3.3.1.3, respectively.

3.3.1.1 Channel Coding

For Internet applications, channel coding is typically used in terms of block codes. Specifically, a video stream is first chopped into segments, each of which is packetized into k packets; then for each segment, a block code (*e.g.*, Tornado code [2]) is applied to the k packets to generate a n-packet block, where $n > k$. Specifically, the channel encoder places the k packets into a group and then creates additional packets from them so that the total number of packets in the group becomes n, where $n > k$ (shown in Fig. 15). This group of packets is transmitted to the receiver, which receives K packets. To perfectly recover a segment, a user must receive K ($K \geq k$) packets in the n-packet block (see Fig. 15). In other words, a user only needs to receive any k packets in the n-packet block so that it can reconstruct all the original k packets.

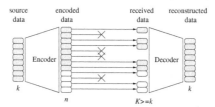

Figure 15: Channel coding/decoding operation.

Since recovery is carried out entirely at the receiver, the channel coding approach can scale to arbitrary number of receivers in a large multicast group. In addition, due to its ability to recover from any k out of n packets regardless of which packets are lost, it allows the network and receivers to discard some of the packets which cannot be handled due to limited bandwidth or processing power. Thus, it is also applicable to heterogeneous networks and receivers with different capabilities. However, there are also some disadvantages associated with channel coding as follows.

1. It increases the transmission rate. This is because channel coding adds $n - k$ redundant packets to every k original packets, which increases the rate by a factor of n/k. In addition, the higher the loss rate is, the higher the transmission rate is required to recover from the loss. The higher the transmission rate is, the more congested the network gets, which leads to an even higher loss rate. This makes channel coding vulnerable for short-term congestion. However, efficiency may be improved by using unequal error protection [2].

2. It increases delay. This is because (1) a channel encoder must wait for all k packets in a segment before it can generate the $n-k$ redundant packets; and (2) the receiver must wait for at least k packets of a block before it can playback the video segment. In addition, recovery from bursty loss requires the use of either longer blocks (*i.e.*, larger k and n) or techniques like interleaving. In either case, delay will be further increased. But for video streaming applications, which can tolerate relatively large delay, the increase in delay may not be an issue.

3. It is not adaptive to varying loss characteristics and it works best only when the packet loss rate is stable. If more than $n - k$ packets of a block are lost, channel coding cannot recover any portion of the original segment. This makes channel coding useless when the short-term loss rate exceeds the recovery capability of the code. On the other hand, if the loss rate is well below the code's recovery capability, the redundant information is more than necessary (a smaller ratio n/k would be more appropriate). To improve the adaptive capability of channel coding, feedback can be used. That is, if the receiver conveys the loss characteristics to the source, the channel encoder can adapt the redundancy accordingly. Note that this requires a closed loop rather than an open loop in the original channel coding design.

A significant portion of previous research on channel coding for video transmission has involved *equal error protection* (EEP), in which all the bits of the compressed video stream are treated equally, and given an equal amount of redundancy. However, the compressed video stream typically does not consist of bits of equal significance. For example, in MPEG, an I-frame is more important than a P-frame while a P-frame is more important than a B-frame. Current research is heavily weighted towards *unequal error protection* (UEP) schemes, in which the more significant information bits are given more protection. A representative work of UEP is the Priority Encoding Transmission (PET) [2]. A key feature of the PET scheme is to allow a user to set different levels (priorities) of error protection for different segments of the video stream. This unequal protection makes PET efficient (less redundancy) and suitable for transporting MPEG video which has an inherent priority hierarchy (*i.e.*, I-, P-, and B-frames).

To provide error recovery in layered multicast video, Tan and Zakhor proposed a receiver-driven hierarchical FEC (HFEC) [127]. In HFEC, additional streams with only FEC redundant information are generated along with the video layers. Each of the FEC streams is used for recovery of a different video layer, and each of the FEC streams is sent to a different multicast group. Subscribing to more FEC groups corresponds to higher level of protection. Like other receiver-driven schemes, HFEC also achieves good trade-off between flexibility of providing recovery and bandwidth efficiency, that is:

- *Flexibility of providing recovery:* Each receiver can independently adjust the desired level of protection based on past reception statistics and the application's delay tolerance.

- *Bandwidth efficiency:* Each receiver will subscribe to only as many redundancy layers as necessary, reducing overall bandwidth utilization.

3.3.1.2 Source Coding-based FEC

Source coding-based FEC (SFEC) is a recently devised variant of FEC for Internet video [16]. Like channel coding, SFEC also adds redundant information to recover from loss. For example, SFEC could add redundant information as follows: the nth packet contains the nth GOB (Group of Blocks) and redundant information about the $(n-1)$th GOB. If the $(n-1)$th packet is lost but the nth packet is received, the receiver can still reconstruct the $(n-1)$th GOB from the redundant information about the $(n-1)$th GOB, which is contained in the nth packet. However, the reconstructed $(n-1)$th GOB has a coarser quality. This is because the redundant information about the $(n-1)$th GOB is a compressed version of the $(n-1)$th GOB with a larger quantizer, resulting in less redundancy added to the nth packet.

The main difference between SFEC and channel coding is the kind of redundant information being added to a compressed video stream. Specifically, channel coding adds redundant information according to a block code (irrelevant to the video) while the redundant information added by SFEC is more compressed versions of the raw video. As a result, when there is packet loss, channel coding could achieve perfect recovery while SFEC recovers the video with reduced quality.

One advantage of SFEC over channel coding is lower delay. This is because each packet can be decoded in SFEC while, under the channel coding approach, both the channel encoder and the channel decoder have to wait for at least k packets of a segment.

Similar to channel coding, the disadvantages of SFEC are: (1) an increase in the transmission rate, and (2) inflexible to varying loss characteristics. However, such inflexibility to varying loss characteristics can also be improved through feedback [16]. That is, if the receiver conveys the loss characteristics to the source, the SFEC encoder can adjust the redundancy accordingly. Note that this requires a closed loop rather than an open loop in the original SFEC coding scheme.

3.3.1.3 Joint Source/Channel Coding

Due to Shannon's separation theorem [115], the coding world was generally divided into two camps: source coding and channel coding. The camp of source coding was concerned with developing efficient source coding techniques while the camp of channel coding was concerned with developing robust channel coding techniques [56]. In other words, the camp of source coding did not take channel coding into account and the camp of channel coding did not consider source coding. However, Shannon's separation theorem is not strictly applicable when the delay is bounded, which is the case for such real-time services as video over the Internet [27]. The motivation of joint source/channel coding for video comes from the following observations:

- **Case A:** According to the rate-distortion theory (shown in Fig. 16(a)) [33], the lower the source-encoding rate R for a video unit, the larger the distortion D of the video unit. That is, $R \downarrow \Rightarrow D \uparrow$.

- **Case B:** Suppose that the total rate (*i.e.*, the source-encoding rate R plus the channel-coding redundancy rate R') is fixed and channel loss characteristics do not change. The higher the source-encoding rate for a video unit is, the lower the channel-coding redundancy rate would be. This leads to a higher probability P_c of the event that the video unit gets corrupted, which translates into a larger distortion of the video unit. That is, $R \uparrow \Rightarrow R' \downarrow \Rightarrow P_c \uparrow \Rightarrow D \uparrow$.

Combining Cases A and B, it can be argued that there exists an optimal source-encoding rate R_o that achieves the minimum distortion D_o (see Fig. 16(b)), given a constant total rate.

The objective of joint source/channel coding is to find the optimal point shown in Fig. 16(b) and design source/channel coding schemes to achieve the optimal point. In other words, finding an optimal point in joint source/channel coding is to make an optimal rate allocation between source coding and channel coding.

Basically, joint source/channel coding is accomplished by three tasks:

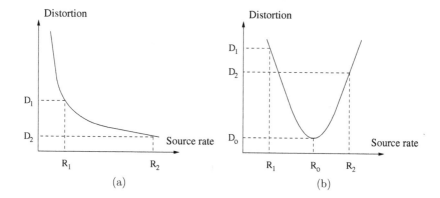

Figure 16: (a) Rate-distortion relation for source coding. (b) Rate-distortion relation for the case of joint source/channel coding with fixed $R + R'$.

- **Task 1:** finding an optimal rate allocation between source coding and channel coding for a given channel loss characteristic,

- **Task 2:** designing a source coding scheme (including specifying the quantizer) to achieve its target rate,

- **Task 3:** designing/choosing channel codes to match the channel loss characteristic and achieve the required robustness.

For the purpose of illustration, Fig. 17 shows an architecture for joint source/channel coding. Under the architecture, a QoS monitor is kept at the receiver side to infer the channel loss characteristics. Such information is conveyed back to the source side through the feedback control protocol. Based on such feedback information, the joint source/channel optimizer makes an optimal rate allocation between the source coding and the channel coding (Task 1) and conveys the optimal rate allocation to the source encoder and the channel encoder. Then the source encoder chooses an appropriate quantizer to achieve its target rate (Task 2) and the channel encoder chooses a suitable channel code to match the channel loss characteristic (Task 3).

An example of joint source/channel coding is the scheme introduced by Davis and Danskin [35] for transmitting images over the Internet. In this scheme, source and channel coding bits are allocated in a way that can minimize an expected distortion measure. As a result, more perceptually important low frequency sub-bands of images are shielded heavily using channel codes

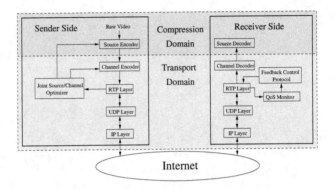

Figure 17: An architecture for joint source/channel coding.

while higher frequencies are shielded lightly. This unequal error protection reduces channel coding overhead, which is most pronounced on bursty channels where a uniform application of channel codes is expensive.

3.3.2 Delay-constrained Retransmission: A Transport Approach

A conventional retransmission scheme, ARQ, works as follows: when packets are lost, the receiver sends feedback to notify the source; then the source retransmits the lost packets. The conventional ARQ is usually dismissed as a method for transporting real-time video since a retransmitted packet arrives at least 3 one-way trip time after the transmission of the original packet, which might exceed the delay required by the application. However, if the one-way trip time is short with respect to the maximum allowable delay, a retransmission-based approach, called delay-constrained retransmission, is a viable option for error control [100, 101].

Typically, one-way trip time is relatively small within the same local area network (LAN). Thus, even delay sensitive interactive video applications could employ delay-constrained retransmission for loss recovery in an LAN environment [37]. Delay-constrained retransmission may also be applicable to streaming video, which can tolerate relatively large delay due to a large receiver buffer and relatively long delay for display. As a result, even in wide area network (WAN), streaming video applications may have sufficient time to recover from lost packets through retransmission and thereby avoid unnecessary degradation in reconstructed video quality.

In the following, we present various delay-constrained retransmission schemes for unicast (Section 3.3.2.1) and multicast (Section 3.3.2.2), respectively.

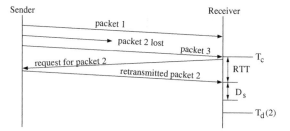

Figure 18: Timing diagram for receiver-based control.

Based on who determines whether to send and/or respond to a retransmission request, we design three delay-constrained retransmission mechanisms for unicast, namely, receiver-based, sender-based, and hybrid control.

Receiver-based control. The objective of the receiver-based control is to minimize the requests of retransmission that will not arrive timely for display. Under the receiver-based control, the receiver executes the following algorithm.

When the receiver detects the loss of packet N:

if $(T_c + RTT + D_s < T_d(N))$

send the request for retransmission of

packet N to the sender;

where T_c is the current time, RTT is an estimated round trip time, D_s is a slack term, and $T_d(N)$ is the time when packet N is scheduled for display. The slack term D_s could include tolerance of error in estimating RTT, the sender's response time to a request, and/or the receiver's processing delay (*e.g.*, decoding). If $T_c + RTT + D_s < T_d(N)$ holds, it is expected that the retransmitted packet will arrive timely for display. The timing diagram for receiver-based control is shown in Fig. 18.

Sender-based control. The objective of the sender-based control is to suppress retransmission of packets that will miss their display time at the receiver. Under the sender-based control, the sender executes the following algorithm.

When the sender receives a request for retransmission

of packet N:

63

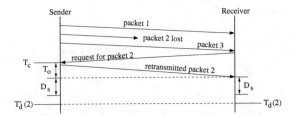

Figure 19: Timing diagram for sender-based control.

$$\text{if } (T_c + T_o + D_s < T'_d(N))$$

retransmit packet N to the receiver

where T_o is the estimated one-way trip time (from the sender to the receiver), and $T'_d(N)$ is an estimate of $T_d(N)$. To obtain $T'_d(N)$, the receiver has to feedback $T_d(N)$ to the sender. Then, based on the differences between the sender's system time and the receiver's system time, the sender can derive $T'_d(N)$. The slack term D_s may include error terms in estimating T_o and $T'_d(N)$, as well as tolerance in the receiver's processing delay (*e.g.*, decoding). If $T_c + RTT + D_s < T'_d(N)$ holds, it can be expected that retransmitted packet will reach the receiver in time for display. The timing diagram for sender-based control is shown in Fig. 19.

Hybrid control. The objective of the hybrid control is to minimize the request of retransmissions that will not arrive for timely display, and to suppress retransmission of the packets that will miss their display time at the receiver. The hybrid control is a simple combination of the sender-based control and the receiver-based control. Specifically, the receiver makes decisions on whether to send retransmission requests while the sender makes decisions on whether to disregard requests for retransmission. The hybrid control could achieve better performance at the cost of higher complexity.

3.3.2.2 Multicast

In the multicast case, retransmission has to be restricted within closely located multicast members. This is because one-way trip times between these members tend to be small, making retransmissions effective in timely recovery. In addition, feedback implosion of retransmission requests is a problem that must be addressed under the retransmission-based approach. Thus, methods are required to limit the number or scope of retransmission requests.

Typically, a logical tree is configured to limit the number/scope of retransmission requests and to achieve local recovery among closely located multicast members [79, 90, 159]. The logical tree can be constructed by statically assigning Designated Receivers (DRs) at each level of the tree to help with retransmission of lost packets [79]. Or it can be dynamically constructed through the protocol used in STructure-Oriented Resilient Multicast (STORM) [159]. By adapting the tree structure to changing network traffic conditions and group memberships, the system could achieve higher probability of receiving retransmission timely.

Similar to the receiver-based control for unicast, receivers in a multicast group can make decisions on whether to send retransmission requests. By suppressing the requests for retransmission of those packets that cannot be recovered in time, bandwidth efficiency can be improved [79]. Besides, using a receiving buffer with appropriate size could not only absorb the jitter but also increase the likelihood of receiving retransmitted packets before their display time [79].

To address heterogeneity problem, a receiver-initiated mechanism for error recovery can be adopted as done in STORM [159]. Under this mechanism, each receiver can dynamically select the best possible DR to achieve good trade-off between desired latency and the degree of reliability.

3.3.3   Error-resilience: A Compression Approach

Error-resilient schemes address loss recovery from the compression perspective. Specifically, they attempt to prevent error propagation or limit the scope of the damage (caused by packet losses) on the compression layer. The standardized error-resilient tools include re-synchronization marking, data partitioning, and data recovery (*e.g.*, reversible variable length codes (RVLC)) [64, 126]. However, re-synchronization marking, data partitioning, and data recovery are targeted at error-prone environments like wireless channels and may not be applicable to the Internet. For Internet video, the boundary of a packet already provides a synchronization point in the variable-length coded bit-stream at the receiver side. On the other hand, since a packet loss may cause the loss of all the motion data and its associated shape/texture data, mechanisms such as re-synchronization marking, data partitioning, and data recovery may not be useful for Internet video applications. Therefore, we do not intend to present the standardized error-resilient tools. Instead, we present two techniques which are promising for robust Internet video transmission, namely, *optimal mode selection* and *multiple description coding*.

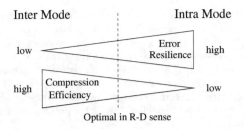

Figure 20: Illustration of optimal mode selection.

3.3.3.1 Optimal Mode Selection

In many video coding schemes, a block, which is a video unit, is coded by reference to a previously coded block so that only the difference between the two blocks needs to be coded, resulting in high coding efficiency. This is called *inter mode*. Constantly referring to previously coded blocks has the danger of error propagation. By occasionally turning off this inter mode, error propagation can be limited. But it will be more costly in bits to code a block all by itself, without any reference to a previously coded block. Such a coding mode is called *intra mode*. Intra-coding can effectively stop error propagation at the cost of compression efficiency while inter-coding can achieve compression efficiency at the risk of error propagation. Therefore, there is a trade-off in selecting a coding mode for each block (see Fig. 20). How to optimally make these choices is the subject of many research investigations [31, 149, 166].

For video communication over a network, a block-based coding algorithm such as H.263 or MPEG-4 [64] usually employs rate control to match the output rate to the available network bandwidth. The objective of rate-controlled compression algorithms is to maximize the video quality under the constraint of a given bit budget. This can be achieved by choosing a mode that minimizes the quantization distortion between the original block and the reconstructed one under a given bit budget [98, 123], which is the so-called R-D optimized mode selection. We refer such R-D optimized mode selection as the classical approach. The classical approach is not able to achieve global optimality under the error-prone environment since it does not consider the network congestion status and the receiver behavior.

To address this problem, an end-to-end approach to R-D optimized mode selection was proposed [149]. Under the end-to-end approach, three factors were identified to have impact on the video presentation quality at the receiver: (1) the source behavior, *e.g.*, quantization and packetization, (2) the path characteristics, and (3) the receiver behavior, *e.g.*, error concealment (see

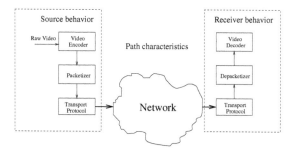

Figure 21: Factors that have impact on the video presentation quality: source behavior, path characteristics, and receiver behavior.

Fig. 21). Based on the characterizations, a theory [149] for globally optimal mode selection was developed. By taking into consideration of the network congestion status and the receiver behavior, the end-to-end approach is shown to be capable of offering superior performance over the classical approach for Internet video applications [149].

3.3.3.2 Multiple Description Coding

Multiple description coding (MDC) is another way to achieve trade-off between compression efficiency and robustness to packet loss [142]. With MDC, a raw video sequence is compressed into multiple streams (referred to as descriptions). Each description provides acceptable visual quality; more combined descriptions provide a better visual quality. The advantages of MDC are: (1) *robustness to loss:* even if a receiver gets only one description (other descriptions being lost), it can still reconstruct video with acceptable quality; and (2) *enhanced quality:* if a receiver gets multiple descriptions, it can combine them together to produce a better reconstruction than that produced from any single description.

However, the advantages do not come for free. To make each description provide acceptable visual quality, each description must carry sufficient information about the original video. This will reduce the compression efficiency compared to conventional single description coding (SDC). In addition, although more combined descriptions provide a better visual quality, a certain degree of correlation between the multiple descriptions has to be embedded in each description, resulting in further reduction of the compression efficiency. Current research effort is to find a good trade-off between the compression efficiency and the reconstruction quality from one description.

3.3.4 Error Concealment: A Compression Approach

When packet loss is detected, the receiver can employ error concealment to conceal the lost data and make the presentation more pleasing to human eyes. Since human eyes can tolerate a certain degree of distortion in video signals, error concealment is a viable technique to handle packet loss [143].

There are two basic approaches for error concealment, namely, spatial and temporal interpolation. In spatial interpolation, missing pixel values are reconstructed using neighboring spatial information, whereas in temporal interpolation, the lost data is reconstructed from data in the previous frames. Typically, spatial interpolation is used to reconstruct missing data in intra-coded frames while temporal interpolation is used to reconstruct missing data in inter-coded frames.

In recent years, numerous error-concealment schemes have been proposed in the literature (refer to [143] for a good survey). Examples include maximally smooth recovery [141], projection onto convex sets [124], and various motion vector and coding mode recovery methods such as motion compensated temporal prediction [50]. However, most error concealment techniques discussed in [143] are only applicable to either ATM or wireless environments, and require substantial additional computation complexity, which is acceptable for decoding still images but not tolerable in decoding real-time video. Therefore, we only describe simple error concealment schemes that are applicable to Internet video communication.

We describe three simple error concealment (EC) schemes as follows.

EC-1: The receiver replaces the whole frame (where some blocks are corrupted due to packet loss) with the previous reconstructed frame.

EC-2: The receiver replaces a corrupted block with the block at the same location from the previous frame.

EC-3: The receiver replaces the corrupted block with the block from the previous frame pointed by a motion vector. The motion vector is copied from its neighboring block when available, otherwise the motion vector is set to zero.

EC-1 and *EC*-2 are special cases of *EC*-3. If the motion vector of the corrupted block is available, *EC*-3 can achieve better performance than *EC*-1 and *EC*-2 while *EC*-1 and *EC*-2 have less complexity than that of *EC*-3.

3.4 Summary

Transporting video over the Internet is an important component of many multimedia applications. Lack of QoS support in the current Internet, and the

Table 1: Taxonomy of the Design Space

		Transport perspective	Compression perspective
Congestion control	Rate control	Source-based	
		Receiver-based	
		Hybrid	
	Rate adaptive encoding		Altering quantizer
			Altering frame rate
	Rate shaping	Selective frame discard	Dynamic rate shaping
Error control	FEC	Channel coding	SFEC
		Joint channel/source coding	
	Delay-constrained retransmission	Sender-based control	
		Receiver-based control	
		Hybrid control	
	Error resilience		Optimal mode selection
			Multiple description coding
	Error concealment		EC-1, EC-2, EC-3

heterogeneity of the networks and end-systems pose many challenging problems for designing video delivery systems. In this chapter, we identified four problems for video delivery systems: bandwidth, delay, loss, and heterogeneity, and surveyed the end system-based approach to addressing these problems.

Over the past several years, extensive research based on the end system-based approach has been conducted and various solutions have been proposed. To depict a big picture, we took a holistic approach from both transport and compression perspectives. With the holistic approach, we presented a framework for transporting real-time Internet video, which consisted of two components: congestion control and error control. We have described various approaches and schemes for the two components. All the possible approaches/schemes for the two components can form a design space. As shown in Table 1, the approaches/schemes in the design space can be classified along two dimensions: the transport perspective and the compression perspective.

To give the reader a clear picture of this design space, we summarize the advantages and disadvantages of the approaches and schemes as follows.

1. Congestion control. There are three mechanisms for congestion control: rate control, rate adaptive video encoding, and rate shaping. Rate control schemes can be classified into three categories: source-based, receiver-based, and hybrid. As shown in Table 2, rate control schemes can follow either the model-based approach or the probe-based approach. Source-based rate control is primarily targeted at unicast and can follow either the model-based approach or the probe-based approach. If applied in multicast, source-based rate control can only follow the probe-based

approach. Source-based rate control needs another component to enforce the rate on the video stream. This component could be either rate adaptive video encoding or rate shaping. Examples of combining source-based rate control with rate adaptive video encoding can be found in [128, 150]. Examples of combining source-based rate control with rate shaping include [66]. Receiver-based and hybrid rate control were proposed to address the heterogeneity problem in multicast video. The advantage of receiver-based control over sender-based control is that the burden of adaptation is moved from the sender to the receivers, resulting in enhanced service flexibility and scalability. Receiver-based rate control can follow either the model-based approach or the probe-based approach. Hybrid rate control combines some of the best features of receiver-based and sender-based control in terms of service flexibility and bandwidth efficiency. But it can only follow the probe-based approach. For video multicast, one advantage of the model-based approach over the probe-based approach is that it does not require exchange of information among the group as is done under the probe-based approach. Therefore, it eliminates processing at each receiver and the bandwidth usage associated with information exchange.

2. Error control. It takes the form of FEC, delay-constrained retransmission, error-resilience or error concealment. There are three kinds of FEC: channel coding, source coding-based FEC, and joint source/channel coding. The advantage of all FEC schemes over TCP is reduction in video transmission latency. Source coding-based FEC can achieve lower delay than channel coding while joint source/channel coding could achieve optimal performance in a rate-distortion sense. The disadvantages of all FEC schemes are: increase in the transmission rate, and inflexibility to varying loss characteristics. A feedback mechanism can be used to improve FEC's inflexibility. Unlike FEC, which adds redundancy to recover from loss that might not occur, a retransmission-based scheme only re-sends the packets that are lost. Thus, a retransmission-based scheme is adaptive to varying loss characteristics, resulting in efficient use of network resources. But delay-constrained retransmission-based schemes may become useless when the round trip time is too large. Optimal mode selection and multiple description coding are two error-resilient mechanisms recently proposed. Optimal mode selection achieves the best trade-off between compression efficiency and error resilience in an R-D sense. The cost of optimal mode selection is its complexity, which is similar to that of motion compensation algorithms. Multiple description coding is another way of trading off compression efficiency with robustness to packet

Table 2: Rate Control

		Model-based approach	Probe-based approach
Rate control	Source-based	Unicast	Unicast/Multicast
	Receiver-based	Multicast	Multicast
	Hybrid		Multicast

loss. The advantage of MDC is its robustness to loss and enhanced quality. The cost of MDC is reduction in compression efficiency. Finally, as the last stage of a video delivery system, error concealment can be used in conjunction with any other techniques (*i.e.*, congestion control, FEC, retransmission, and error-resilience).

This chapter surveyed general end-system-based techniques and these techniques are also applicable (with modifications if necessary) to video transport over networks with wireless segments.

CHAPTER 4

EFFECTIVE CAPACITY CHANNEL MODELING

From this chapter through Chapter 7, we will be concerned with network-centric QoS provisioning.

4.1 Introduction

To provide explicit QoS guarantees such as a data rate, delay bound, and delay-bound violation probability triplet, it is necessary to analyze a QoS provisioning system in terms of these QoS measures. This task requires characterization of the service (channel modeling), and queueing analysis of the system. Specifically, a general methodology of designing QoS provisioning mechanisms at a network node, involves four steps:

1. Channel measurement: *e.g.*, measure the channel capacity process [71]. It requires channel estimation at the receiver side and feedback of channel estimates to the transmitter.

2. Channel modeling: *e.g.*, use a Markov-modulated Poisson process to model the channel capacity process [71].

3. Deriving QoS measures: *e.g.*, analyze the queue of the system and derive the delay distribution, given the Markov-modulated Poisson process as the service model and assuming a certain Markovian traffic model [71].

4. Relating the control parameters of QoS provisioning mechanisms to the derived QoS measures: *e.g.*, relate the amount of allocated resource to the QoS measures. If such a relationship is known, given the QoS requirements specified by a user, we can calculate how much resource needs to be allocated to satisfy the QoS.

Steps 1 to 3 are intended to analyze the QoS provisioning mechanisms, whereas step 4 is aimed at designing the QoS provisioning mechanisms.

However, the main obstacle of applying the four steps in QoS provisioning, is *high complexity* in characterizing the relation between the control parameters and the calculated QoS measures, based on existing channel models, *i.e.*, physical-layer channel models (refer to Section 2.5). This is because the physical-layer channel models (*e.g.*, Rayleigh fading model with a specified Doppler spectrum) do not explicitly characterize a wireless channel in terms of the link-level QoS metrics specified by users, such as data rate, delay and delay-violation probability. To use the physical-layer channel models for QoS support, we first need to estimate the parameters for the channel model, and then extract the link-level QoS metrics from the model. This two-step approach is obviously complex, and may lead to inaccuracies due to possible approximations in extracting QoS metrics from the models.

Recognizing that the limitation of physical-layer channel models in QoS support, is the difficulty in analyzing queues using them, we propose moving the channel model up the protocol stack, from the physical-layer to the link-layer. We call the resulting model an *effective capacity* (EC) channel model, because it captures a generalized link-level capacity notion of the fading channel. Figure 4 (in Section 2.5) illustrates the difference between the conventional physical-layer channels and the link-layer channel.[1] In this chapter, we consider small-scale fading [109] in the radio-layer channel.

To summarize, the effective capacity channel model that we propose, aims to characterize wireless channels in terms of functions that can be easily mapped to link-level QoS metrics, such as delay-bound violation probability. Furthermore, we propose a novel channel estimation algorithm that allows practical and accurate measurements of the effective capacity model functions. The EC model captures the effect of channel fading on the queueing behavior of the link, using a computationally simple yet accurate model, and thus, is a critical tool for designing efficient QoS provisioning mechanisms.

The remainder of this chapter is organized as follows. In Section 4.2, we elaborate on the QoS guarantees that motivate us to search for a link-layer model. We describe usage parameter control (UPC) traffic characterization, and its dual, the service curve (SC) network service characterization. We show that these concepts, borrowed from networking literature, lead us to consider the effective capacity model of wireless channels. In Section 4.3, we formally define the effective capacity channel model, in terms of two functions, probability of non-empty buffer and QoS exponent. We then describe an estimation

---

[1] In Figure 4, we use Shannon's channel capacity to represent the instantaneous channel capacity. In practical situations, the instantaneous channel capacity is $\log(1 + SNR/\Gamma_{link})$, where Γ_{link} is determined by the modulation scheme and the channel code used.

algorithm, which accurately estimates these functions, with very low complexity. Section 4.4 shows simulation results that demonstrate the advantage of using the EC channel model to accurately predict QoS, under a variety of conditions. This leads to efficient admission control and resource reservation. In Section 4.5, we extend our EC model to the case of frequency selective fading channels. In Section 4.6, we use the effective capacity technique to derive QoS measures for more general situations, *i.e.*, networks with multiple wireless links, variable-bit-rate sources, packetized traffic, and wireless channels with non-negligible propagation delay. Section 4.7 summarizes this chapter. Table 3 lists the notations used in the rest of the book.

4.2 Motivation for Using Link-layer Channel Models

Physical-layer channel models have been extremely successful in wireless transmitter/receiver design, since they can be used to predict physical-layer performance characteristics such as bit/frame error rates as a function of SNR. These are very useful for circuit switched applications, such as cellular telephony. However, future wireless systems will need to handle increasingly diverse multimedia traffic, which are expected to be primarily packet switched. For example, the new Wideband Code Division Multiple Access (W-CDMA) specifications make explicit provisions for 3G networks to evolve over time, from circuit switching to packet switching. The key difference between circuit switching and packet switching, from a link-layer design viewpoint, is that packet switching requires *queueing* analysis of the link. Thus, it becomes important to characterize the effect of the data traffic pattern, as well as the channel behavior, on the performance of the communication system.

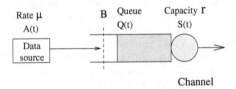

Figure 22: A queueing system model.

QoS guarantees have been heavily researched in the *wired* networks (*e.g.*, Asynchronous Transfer Mode (ATM) and Internet Protocol (IP) networks). These guarantees rely on the queueing model shown in Figure 22. This figure shows that the source traffic and the network service are matched using

$Pr\{\cdot\}$	:	probability of the event $\{\cdot\}$.
$\Gamma(t)$	:	a traffic envelope.
$\Psi(t)$	:	a network service curve.
$A(t)$	:	the amount of source data over the time interval $[0, t)$.
$S(t)$	:	the actual service of a channel in bits, over the time interval $[0, t)$.
$r(t)$	:	the instantaneous capacity of a channel at time t.
$\tilde{S}(t)$	:	the service provided by a channel, $i.e.$, $S(t) = \int_0^t r(\tau)\mathrm{d}\tau$.
$\lambda_p^{(s)}$	:	the peak rate of a source.
$\lambda_s^{(s)}$	:	the sustainable rate of a source.
$\sigma^{(s)}$	:	the leaky-bucket size for the source model.
$\lambda_s^{(c)}$	:	the channel sustainable rate.
$\sigma^{(c)}$	:	the maximum fade duration of a channel.
r	:	the service rate of a queue.
B	:	the buffer size of a queue.
$\Lambda(u)$	:	the asymptotic log-moment generating function of a stochastic process.
$\alpha^{(s)}(u)$	:	the effective bandwidth of a source.
$\alpha(u)$	:	the effective capacity of a channel.
$Q(t)$	:	the length of a queue at time t.
$D(t)$	:	the delay experienced by a packet arriving at time t.
D_{max}	:	the delay bound required by a connection.
ε	:	the target QoS violation probability for a connection.
θ	:	the QoS exponent of a connection.
γ	:	probability of the event that a queue is non-empty.
$S(f)$	:	the Doppler spectrum (power spectral density) of a channel.
f_m	:	the maximum Doppler frequency for a mobile terminal.
f_c	:	the carrier frequency.
$det(.)$	:	the determinant of a matrix.
r_{awgn}	:	the capacity of an additive white Gaussian noise (AWGN) channel.

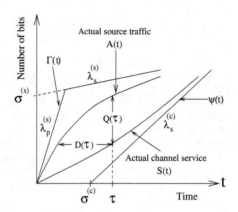

Figure 23: Traffic and service characterization.

a First-In-First-Out (FIFO) buffer (queue). Thus, the queue prevents loss of packets that could occur when the source rate is more than the service rate, at the expense of increasing the delay. Queueing analysis, which is needed to design appropriate admission control and resource reservation algorithms [5, 116], requires source *traffic characterization* and *service characterization*. The most widely used approach for traffic characterization, is to require that the amount of data (*i.e.*, bits as a function of time t) produced by a source conform to an upper bound, called the *traffic envelope* $\Gamma(t)$. The service characterization for guaranteed service is a guarantee of a minimum service (*i.e.*, bits communicated as a function of time) level, specified by a *service curve* $\Psi(t)$ [54]. Functions $\Gamma(t)$ and $\Psi(t)$ are specified in terms of certain traffic and service parameters respectively. Examples include the UPC parameters used in ATM [5] for traffic characterization, and the traffic specification T-SPEC and the service specification R-SPEC fields used with the resource reservation protocol (RSVP) [20, 54] in IP networks.

To elaborate on this point, a traffic envelope $\Gamma(t)$ characterizes the source behavior in the following manner: over any window of size t, the amount of actual source traffic $A(t)$ does not exceed $\Gamma(t)$ (see Figure 23). For example, the UPC parameters specify $\Gamma(t)$ by,

$$\Gamma(t) = \min\{\lambda_p^{(s)}t, \lambda_s^{(s)}t + \sigma^{(s)}\} \qquad (16)$$

76

where $\lambda_p^{(s)}$ is the peak data rate, $\lambda_s^{(s)}$ the sustainable rate, and $\sigma^{(s)}$ the leaky-bucket size [54]. As shown in Figure 23, the curve $\Gamma(t)$ consists of two segments; the first segment has a slope equal to the peak source data rate $\lambda_p^{(s)}$, while the second segment, has a slope equal to the sustainable rate $\lambda_s^{(s)}$, with $\lambda_s^{(s)} < \lambda_p^{(s)}$. $\sigma^{(s)}$ is the Y-axis intercept of the second segment. $\Gamma(t)$ has the property that $A(t) \leq \Gamma(t)$ for any time t.

Just as $\Gamma(t)$ upper bounds the source traffic, a network service curve $\Psi(t)$ lower bounds the actual service $S(t)$ that a source will receive. $\Psi(t)$ has the property that $\Psi(t) \leq S(t)$ for any time t. Both $\Gamma(t)$ and $\Psi(t)$ are negotiated during the admission control and resource reservation phase. An example of a network service curve is the R-SPEC curve used for guaranteed service in IP networks,

$$\Psi(t) = [\lambda_s^{(c)}(t - \sigma^{(c)})]^+ \tag{17}$$

where $[x]^+ = \max\{x, 0\}$, $\lambda_s^{(c)}$ is the constant service rate and $\sigma^{(c)}$ the delay error term (due to propagation delay, link sharing and so on). This curve is illustrated in Figure 23. $\Psi(t)$ consists of two segments; the horizontal segment indicates that no packet is being serviced due to propagation delay, etc., for a time interval equal to the delay error term $\sigma^{(c)}$, while the second segment has a slope equal to the service rate $\lambda_s^{(c)}$. In the figure, we also observe that (1) the horizontal difference between $A(t)$ and $S(t)$, denoted by $D(\tau)$, is the delay experienced by a packet departing at time τ; (2) the vertical difference between the two curves, denoted by $Q(\tau)$, is the queue length built up at time τ, due to packets that have not been served yet.

In contrast to packet-switched wireline networks, providing QoS guarantees in packet-switched wireless networks is a challenging problem. This is because wireless channels have low reliability, and time varying capacities, which may cause severe QoS violations. Unlike wireline links, which typically have a constant capacity, the capacity of a wireless channel depends upon such random factors as multipath fading, co-channel interference, and noise disturbances. Consequently, providing QoS guarantees over wireless channels requires accurate models of their *time-varying capacity*, and effective utilization of these models for QoS support.

The simplicity of the service curves discussed earlier motivates us to define the time-varying capacity of a wireless channel as in (17). Specifically, we hope to lower bound the channel service using two parameters, the channel sustainable rate $\lambda_s^{(c)}$, and the maximum fade duration $\sigma^{(c)}$.[2] However, physical-layer

---

[2] $\lambda_s^{(c)}$ and $\sigma^{(c)}$ are meant to be in a statistical sense. The maximum fade duration $\sigma^{(c)}$ is a parameter that relates the delay constraint to the channel service; it determines the probability $\sup_t Pr\{S(t) < \Psi(t)\}$. We will see later that $\sigma^{(c)}$ is specified by the source with

wireless channel models do not explicitly characterize the channel in terms of such link-layer QoS metrics as data rate, delay and delay-violation probability. For this reason, we are forced to look for alternative channel models.

A tricky issue that surfaces, is that a wireless channel has a capacity that varies *randomly* with time. Thus, an attempt to provide a strict lower bound (*i.e.*, the deterministic service curve $\Psi(t)$, used in IP networks) will most likely result in extremely conservative guarantees. For example, in a Rayleigh or Ricean fading channel, the only lower bound that can be *deterministically* guaranteed is a capacity[3] of zero! This conservative guarantee is clearly useless. Therefore, we propose to extend the concept of deterministic service curve $\Psi(t)$, to a *statistical* version, specified as the pair $\{\Psi(t), \varepsilon\}$. The statistical service curve $\{\Psi(t), \varepsilon\}$ specifies that the service provided by the channel, denoted as $\tilde{S}(t)$, will always satisfy the property that $\sup_t Pr\{\tilde{S}(t) < \Psi(t)\} \leq \varepsilon$. In other words, ε is the probability that the wireless channel will not be able to support the pledged service curve $\Psi(t)$. For most practical values of ε, a *non-zero* service curve $\Psi(t)$ can be guaranteed.

To summarize, we propose to extend the QoS mechanisms used in wired networks to wireless links, by using the traffic and service characterizations popularly used in wired networks; namely the traffic envelope $\Gamma(t)$ and the service curve $\Psi(t)$ respectively. However, recognizing that the time-varying wireless channel cannot deterministically guarantee a useful service curve, we propose to use a statistical service curve $\{\Psi(t), \varepsilon\}$.

As mentioned earlier, it is hard to extract a statistical service curve using the existing physical-layer channel models. In fact, in Section 4.3.4, we show how physical-layer channel models can be used to derive $\{\Psi(t), \varepsilon\}$, in an integral form. There, the reader will see that 1) it is not always possible to extract $\{\Psi(t), \varepsilon\}$ from the physical-layer model (such as, when only the Doppler spectrum, but not the higher-order statistics are known), and 2) even if it is possible, the computation involved may make the extraction extremely hard to implement. This motivates us to consider link-layer modeling, which we describe in Section 4.3. The philosophy here is that, we want to model the wireless channel at the layer in which we intend to use the model.

$\sigma^{(c)} = D_{max}$, where D_{max} is the delay bound required by the source.

[3]The capacity here is meant to be delay-limited capacity, which is the maximum rate achievable with a prescribed delay bound (see [57] for details).

4.3 Effective Capacity Model of Wireless Channels

Section 4.2 argued that QoS guarantees can be achieved if a statistical service curve can be calculated for the given wireless channel. Thus, we need to calculate a service curve $\Psi(t)$, such that for a given $\varepsilon > 0$, the following probability bound on the channel service $\tilde{S}(t)$ is satisfied,

$$\sup_t Pr\{\tilde{S}(t) < \Psi(t)\} \le \varepsilon \qquad (18)$$

Further, $\Psi(t)$ is restricted to being specified by the parameters $\{\lambda_s^{(c)}, \sigma^{(c)}\}$, as below ((17), which we reproduce, for convenience),

$$\Psi(t) = [\lambda_s^{(c)}(t - \sigma^{(c)})]^+ \qquad (19)$$

Therefore, the statistical service curve specification requires that we relate its parameters $\{\lambda_s^{(c)}, \sigma^{(c)}, \varepsilon\}$ to the fading wireless channel. Note that a (non-fading) AWGN channel of capacity r_{awgn} can be specified by the triplet $\{r_{awgn}, 0, 0\}$. i.e., an AWGN channel can guarantee constant data rate.

At first sight, relating $\{\lambda_s^{(c)}, \sigma^{(c)}, \varepsilon\}$ to the fading wireless channel behavior seems to be a hard problem. However, at this point, we use the idea that the service curve $\Psi(t)$ is a *dual* of the traffic envelope $\Gamma(t)$. A rich body of literature exists on the so-called *theory of effective bandwidth* [23], which models the statistical behavior of *traffic*. In particular, the theory shows that the relation

$$\sup_t Pr\{Q(t) \ge B\} \le \varepsilon \qquad (20)$$

is satisfied for large B, by choosing two parameters (which are functions of the channel rate r) that depend on the actual data traffic; namely, the probability of non-empty buffer, and the effective bandwidth of the source. *Thus, a source model defined by these two functions fully characterizes the source from a QoS viewpoint.* The duality between (18) and (20) indicates that it may be possible to adapt the theory of effective bandwidth to service curve characterization. This adaptation will point to a new channel model, which we call the *effective capacity (EC) channel model.* Thus, the EC channel model can be thought of as the dual of the effective bandwidth source model, which is commonly used in networking.

The rest of this section is organized as follows. In Section 4.3.1, we present the theory of effective bandwidth using the framework of Chang and Thomas [23]. An accurate and efficient source traffic estimation algorithm exists [88], which can be used to estimate the functions of the *effective bandwidth source model*. Therefore, we use a dual estimation algorithm to estimate the functions of the proposed *effective capacity channel* model in Section 4.3.2. In Section 4.3.3, we provide physical interpretation of our channel model. Section 4.3.4 shows that in the special case of *Rayleigh fading channel at low SNRs*, it is possible to extract the service curve from a physical-layer channel model. For Rayleigh fading channels at high SNRs, the extraction is complicated, whereas the extraction may not even be possible for other types of fading. Therefore, our link-layer EC model has substantial advantage over physical-layer models, in specifying service curves, and hence QoS.

4.3.1 Theory of Effective Bandwidth

For convenience, we reproduce the definition of effective bandwidth presented in Section 2.2.1. The stochastic behavior of a source traffic process can be modeled asymptotically by its effective bandwidth. Consider an arrival process $\{A(t),\ t \geq 0\}$ where $A(t)$ represents the amount of source data (in bits) over the time interval $[0,\ t)$. Assume that the asymptotic log-moment generating function of a stationary process $A(t)$, defined as

$$\Lambda(u) = \lim_{t \to \infty} \frac{1}{t} \log E[e^{uA(t)}], \tag{21}$$

exists for all $u \geq 0$. Then, the *effective bandwidth function* of $A(t)$ is defined as

$$\alpha^{(s)}(u) = \frac{\Lambda(u)}{u} \qquad ,\ \forall\ u > 0. \tag{22}$$

Consider a queue of infinite buffer size served by a channel of *constant* service rate r (see Figure 22), such as an AWGN channel. Due to the possible mismatch between $A(t)$ and $S(t)$, the queue length $Q(t)$ (see Figure 23) could be non-zero. Denote $D(\infty)$ the steady state of $Q(t)$. Using the theory of large deviations, it can be shown that the probability of $Q(\infty)$ exceeding a threshold B satisfies [23, 24]

$$Pr\{Q(\infty) \geq B\} \sim e^{-\theta_B(r)B} \qquad \text{as } B \to \infty, \tag{23}$$

where $f(x) \sim g(x)$ means that $\lim_{x \to \infty} f(x)/g(x) = 1$. However, it is found that for smaller values of B, the following approximation is more accurate [28]

$$Pr\{Q(\infty) \geq B\} \approx \gamma^{(s)}(r)e^{-\theta_B(r)B}, \tag{24}$$

where both $\gamma^{(s)}(r)$ and $\theta_B(r)$ are functions of channel capacity r. According to the theory, $\gamma^{(s)}(r) = Pr\{Q(\infty) \geq 0\}$ is the *probability that the buffer is non-empty*, while the *QoS exponent* θ_B is the solution of $\alpha^{(s)}(\theta_B) = r$. Thus, the pair of functions $\{\gamma^{(s)}(r), \theta_B(r)\}$ model the source. Note that $\theta_B(r)$ is simply the inverse function corresponding to the effective bandwidth function $\alpha^{(s)}(u)$.

If the quantity of interest is the delay $D(t)$ experienced by a source packet departing at time t (see Figure 23), then the probability of $D(t)$ exceeding a delay bound D_{max} satisfies

$$\sup_t Pr\{D(t) \geq D_{max}\} = Pr\{D(\infty) \geq D_{max}\} \approx \gamma^{(s)}(r)e^{-\theta^{(s)}(r)D_{max}}, \quad (25)$$

where $D(\infty)$ denotes the steady state of $D(t)$. Note that $\theta^{(s)}(r)$ in (25) is different from $\theta_B(r)$ in (24). The relationship between them is $\theta^{(s)}(r) = \theta_B(r) \times r$ [167, page 57].

Thus, the key point is that, for a source modeled by the pair $\{\gamma^{(s)}(r), \theta^{(s)}(r)\}$, which has a communication delay bound of D_{max}, and can tolerate a delay-bound violation probability of at most ε, the effective bandwidth concept shows that the constant channel capacity should be at least r, where r is the solution to $\varepsilon = \gamma^{(s)}(r)e^{-\theta^{(s)}(r)D_{max}}$. In terms of the traffic envelope $\Gamma(t)$ (Figure 23), the slope $\lambda_s^{(s)} = r$ and $\sigma^{(s)} = rD_{max}$.

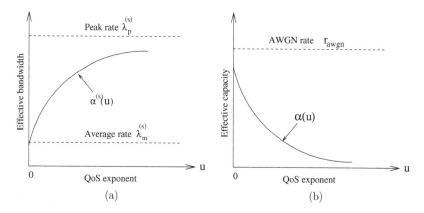

Figure 24: (a) Effective bandwidth function $\alpha^{(s)}(u)$ and (b) Effective capacity function $\alpha(u)$.

Figure 24(a) shows a typical effective bandwidth function. It can be easily proved that $\alpha^{(s)}(0)$ is equal to the average data rate of the source, while $\alpha^{(s)}(\infty)$ is equal to the peak data rate. From (25), note that a source that has a more stringent QoS requirement (*i.e.*, smaller D_{max} or smaller ε), will need a larger QoS exponent $\theta^{(s)}(r)$.

Ref. [88] shows a simple and efficient algorithm to estimate the source model functions $\gamma^{(s)}(r)$ and $\theta^{(s)}(r)$. In the following section, we use the duality between traffic modeling ($\{\gamma^{(s)}(r), \theta^{(s)}(r)\}$), and channel modeling to propose an effective capacity channel model, specified by a pair of functions $\{\gamma(\mu), \theta(\mu)\}$. It is clear that we intend $\{\gamma(\mu), \theta(\mu)\}$ to be the channel duals of the source functions $\{\gamma^{(s)}(r), \theta^{(s)}(r)\}$. Just as the constant *channel rate* r is used in source traffic modeling, we use the constant *source traffic rate* μ in modeling the channel. Furthermore, we adapt the source estimation algorithm in [88] to estimate the channel model functions $\{\gamma(\mu), \theta(\mu)\}$.

4.3.2 Effective Capacity Channel Model

Let $r(t)$ be the instantaneous channel capacity at time t. Define $\tilde{S}(t) = \int_0^t r(\tau)\mathrm{d}\tau$, which is the service provided by the channel. Note that the channel service $\tilde{S}(t)$ is different from the actual service $S(t)$ received by the source; $\tilde{S}(t)$ only depends on the instantaneous channel capacity and thus is independent of the arrival $A(t)$. Paralleling the development in Section 4.3.1, we assume that,

$$\Lambda(-u) = \lim_{t \to \infty} \frac{1}{t} \log E[e^{-u\tilde{S}(t)}] \qquad (26)$$

exists for all $u \geq 0$. This assumption is valid, for example, for a stationary Markov fading process $r(t)$. Then, the *effective capacity function* of $r(t)$ is defined as

$$\alpha(u) = \frac{-\Lambda(-u)}{u} \qquad , \; \forall \, u > 0. \qquad (27)$$

That is,

$$\alpha(u) = -\lim_{t \to \infty} \frac{1}{ut} \log E[e^{-u \int_0^t r(\tau)\mathrm{d}\tau}], \; \forall \, u > 0. \qquad (28)$$

Consider a queue of infinite buffer size supplied by a data source of *constant* data rate μ (see Figure 22). The theory of effective bandwidth presented in Section 4.3.1 can be easily adapted to this case. The difference is that whereas in Section 4.3.1, the source rate was variable while the channel capacity was constant, in this section, the source rate is constant while the channel capacity is variable. Similar to (25), it can be shown that the probability of $D(t)$ exceeding a delay bound D_{max} satisfies

$$\sup_t Pr\{D(t) \geq D_{max}\} = Pr\{D(\infty) \geq D_{max}\} \approx \gamma(\mu)e^{-\theta(\mu)D_{max}}. \qquad (29)$$

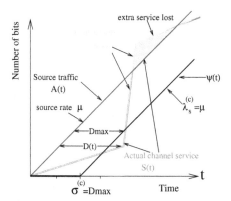

Figure 25: Relation between service curve and delay-bound violation.

where $\{\gamma(\mu), \theta(\mu)\}$ are functions of source rate μ. The approximation in (29) is accurate for large D_{max}, but we will show later, in the simulations, that this approximation is also accurate even for smaller values of D_{max}.

For a given source rate μ, $\gamma(\mu) = Pr\{Q(\infty) \geq 0\}$ is again the *probability that the buffer is non-empty*, while the *QoS exponent* $\theta(\mu)$ is defined as

$$\theta(\mu) = \mu\alpha^{-1}(\mu), \tag{30}$$

where $\alpha^{-1}(\cdot)$ is the inverse function of $\alpha(u)$. Thus, the pair of functions $\{\gamma(\mu), \theta(\mu)\}$ model the channel.

So, when using a link that is modeled by the pair $\{\gamma(\mu), \theta(\mu)\}$, a source that requires a communication delay bound of D_{max}, and can tolerate a delay-bound violation probability of at most ε, needs to limit its data rate to a maximum of μ, where μ is the solution to $\varepsilon = \gamma(\mu)e^{-\theta(\mu)D_{max}}$. In terms of the service curve $\Psi(t)$ (Figure 23), the channel sustainable rate $\lambda_s^{(c)} = \mu$ and $\sigma^{(c)} = D_{max}$. It is clear that when the buffer is empty, the extra service $(\tilde{S}(t) - S(t))$ will be lost, *i.e.*, not used for data transmission (see Figure 25). So, we have $\tilde{S}(t) \geq S(t)$ for any time t. Furthermore, we know that the event $\{D(t) > D_{max}\}$ and the event $\{S(t) < \Psi(t)\}$ are the same. This can be illustrated by Figure 25: whenever the curve $S(t)$ is below $\Psi(t)$, the horizontal line $D(t)$ will cross the line $\Psi(t)$, *i.e.*, we have an event $\{D(t) > D_{max}\}$, since the horizontal distance between $A(t)$ and $\Psi(t)$ is D_{max}. Then, we have

$$\sup_t Pr\{D(t) > D_{max}\} \overset{(a)}{=} \sup_t Pr\{S(t) < \Psi(t)\}$$

$$\overset{(b)}{\geq} \sup_t Pr\{\tilde{S}(t) < \Psi(t)\} \qquad (31)$$

where (a) follows from the fact that $\{D(t) > D_{max}\}$ and $\{S(t) < \Psi(t)\}$ are the same event, and (b) from the fact that $\tilde{S}(t) \geq S(t)$ for any t. From (31), it can be seen that $\sup_t Pr\{D(t) > D_{max}\} \leq \varepsilon$ implies $\sup_t Pr\{\tilde{S}(t) < \Psi(t)\} \leq \varepsilon$.

Figure 24(b) shows that the effective capacity $\alpha(u)$ decreases with increasing QoS exponent u; that is, as the QoS requirement becomes more stringent, the source rate that a wireless channel can support with this QoS guarantee, decreases. The channel sustainable rate $\lambda_s^{(c)}$ is upper bounded by the AWGN capacity r_{awgn}, and lower bounded by the minimum rate 0. Figures 24(a) and 24(b) together illustrate the duality of the effective bandwidth source model and the effective capacity channel model.

The function pair $\{\gamma(\mu), \theta(\mu)\}$ defines our proposed effective capacity channel model. The definition of these functions shows that the EC model is a link-layer model, because it directly characterizes the queueing behavior at the link-layer. From (29), it is clear that the QoS metric can be easily extracted from the EC channel model.

Now that we have shown that QoS metric calculation is trivial, once the EC channel model is known, we need to specify a simple (and hopefully, accurate) channel estimation algorithm. Such an algorithm should estimate the functions $\{\gamma(\mu), \theta(\mu)\}$ from channel measurements, such as the measured SNR or channel capacity $r(t)$.

Let us take a moment to think about how we could use the existing physical-layer channel models to estimate $\{\gamma(\mu), \theta(\mu)\}$. An obvious fact that emerges is that if a channel model specifies only the *marginal* Probability Density Function (PDF) at any time t (such as Ricean PDF), along with the Doppler spectrum (such as Gans Doppler spectrum), which is *second order statistics*, then the model does not have enough information to calculate the effective capacity function (27)! Indeed, such a calculation would need higher order joint statistics, which cannot be obtained merely from the Doppler spectrum. Thus, only approximations can be made in this case. For a Rayleigh fading distribution, the joint PDF of channel gains will be complex Gaussian, and hence the Doppler spectrum is enough to calculate the effective capacity. This result is presented in Section 4.3.4. However, it will be shown there, that even in the Rayleigh fading case, the calculation is complicated, and therefore, not likely to be practical.

84

Assume that the channel fading process $r(t)$ is stationary and ergodic. Then, a simple algorithm to estimate the functions $\{\gamma(\mu), \theta(\mu)\}$ is the following (adapted from [42, 88]),

$$\frac{\gamma(\mu)}{\theta(\mu)} = E[D(\infty)] \qquad (32)$$

$$= \tau_s(\mu) + \frac{E[Q(\infty)]}{\mu} \qquad , \text{ and} \qquad (33)$$

$$\gamma(\mu) = Pr\{D(\infty) > 0\} \qquad (34)$$

where $\tau_s(\mu)$ is the average remaining service time of a packet being served. Note that $\tau_s(\mu)$ is zero for a fluid model (assuming infinitesimal packet size). The intuition in (32) is that, since the distribution of $D(\infty)$ is approximately exponential for large D (see (29)), then $E[D(\infty)]$ is given by (32). Now, the delay $D(\infty)$ is the sum of the delay incurred due to the packet already in service, and the delay in waiting for the queue $Q(\infty)$ to clear. This results in equation (33), using Little's theorem. Substituting $D_{max} = 0$ in (29) results in (34).

Solving (33) for $\theta(\mu)$, we obtain,

$$\theta(\mu) = \frac{\gamma(\mu) \times \mu}{\mu \times \tau_s(\mu) + E[Q(\infty)]}. \qquad (35)$$

Eqs. (34) and (35) show that the functions γ and θ can be estimated by estimating $Pr\{D(\infty) > 0\}$, $\tau_s(\mu)$, and $E[Q(\infty)]$. The latter can be estimated by taking a number of samples, say N_T, over an interval of length T, and recording the following quantities at the n-th sampling epoch ($n = 1, 2, \cdots, N_T$): $S(n)$ the indicator of whether a packets is in service ($S(n) \in \{0, 1\}$), $Q(n)$ the number of bits in the queue (excluding the packet in service), and $\tau(n)$ the remaining service time of the packet in service (if there is one in service). The following sample means are computed,

$$\hat{\gamma} = \frac{1}{N_T} \sum_{t=1}^{N_T} S(n), \qquad (36)$$

$$\hat{q} = \frac{1}{N_T} \sum_{t=1}^{N_T} Q(n). \qquad (37)$$

and

$$\hat{\tau}_s = \frac{1}{N_T} \sum_{t=1}^{N_T} \tau(n). \qquad (38)$$

85

Then, from Eq. (35), we have,

$$\hat{\theta} = \frac{\hat{\gamma} \times \mu}{\mu \times \hat{\tau}_s + \hat{q}} \tag{39}$$

Eqs. (36) through (39) constitute our channel estimation algorithm, to estimate the EC channel model functions $\{\gamma(\mu), \theta(\mu)\}$. They can be used to predict the QoS by approximating Eq. (29) with

$$Pr\{D(\infty) \geq D_{max}\} \approx \hat{\gamma} e^{-\hat{\theta} D_{max}}. \tag{40}$$

Furthermore, if the ultimate objective of EC channel modeling is to compute an appropriate service curve $\Psi(t)$, then as mentioned earlier, given the delay bound D_{max} and the target delay-bound violation probability ε of a connection, we can find $\Psi(t) = \{\sigma^{(c)}, \lambda_s^{(c)}\}$ by, 1) setting $\sigma^{(c)} = D_{max}$, 2) solving Eq. (40) for μ and setting $\lambda_s^{(c)} = \mu$. A fast binary search procedure that estimates $\lambda_s^{(c)}$ for a given D_{max} and ε, is shown in Appendix A.1.

This section introduced the effective capacity channel model, which is parameterized by the pair of functions $\{\gamma(\mu), \theta(\mu)\}$. It was shown that these functions can be easily used to derive QoS guarantees (29), such as a bound that uses $\{D_{max}, \varepsilon\}$. Furthermore, this section specified a simple and efficient algorithm ((36) through (39)) to estimate $\{\gamma(\mu), \theta(\mu)\}$, which can then be used in (29). This completes the specification of our link-layer model.

The EC channel model and its application are summarized below.

1. $\{\gamma(\mu), \theta(\mu)\}$ is the EC channel model, which exists if the log-moment generating function $\Lambda(-u)$ in (26) exists (*e.g.*, for a stationary Markov fading process $r(t)$).

2. In addition to its stationarity, if $r(t)$ is also ergodic, then $\{\gamma(\mu), \theta(\mu)\}$ can be estimated by Eqs. (36) through (39).

3. Given the EC channel model, the QoS $\{\mu, D_{max}, \varepsilon\}$ can be computed by Eq. (40), where $\varepsilon = Pr\{D(\infty) \geq D_{max}\}$.

4. The resulting QoS $\{\mu, D_{max}, \varepsilon\}$ corresponds directly to the service curve specification $\{\lambda_s^{(c)}, \sigma^{(c)}, \varepsilon'\}$ with $\lambda_s^{(c)} = \mu$, $\sigma^{(c)} = D_{max}$, and $\varepsilon' \leq \varepsilon$.

4.3.3 Physical Interpretation of Our Model $\{\gamma(\mu), \theta(\mu)\}$

We stress that the model presented in the previous section, $\{\gamma(\mu), \theta(\mu)\}$, is not just a result of mathematics (*i.e.*, large deviation theory). But rather, the model has direct physical interpretation, *i.e.*, $\{\gamma(\mu), \theta(\mu)\}$ corresponds to marginal Cumulative Distribution Function (CDF) and Doppler spectrum of the underlying physical-layer channel. This correspondence can be illustrated as follows.

- The probability of non-empty buffer, $\gamma(\mu)$, is similar to the concept of marginal CDF (*e.g.*, Rayleigh/Ricean distribution), or equivalently, outage probability (the probability that the received SNR falls below a certain specified threshold). As shown later in Figure 29, different marginal CDF of the underlying physical-layer channel, corresponds to different $\gamma(\mu)$. However, the two functions, marginal CDF (*i.e.*, outage probability) and $\gamma(\mu)$, are not equal. The reason is that the probability of non-empty buffer takes into account the effect of packet accumulation in the buffer, while the outage probability does not (*i.e.*, an arrival packet will be immediately discarded if the SNR falls below a threshold). Therefore, the probability of non-empty buffer is larger than the outage probability, because buffering causes longer busy periods, compared with the non-buffered case.

  From Figure 29, we observe that $\gamma(\mu)$ and marginal CDF have similar behavior, *i.e.*, 1) both increases with the source rate μ; 2) a large outage probability at the physical layer results in a large $\gamma(\mu)$ at the link layer. Thus, $\gamma(\mu)$ does reflect the marginal CDF of the underlying wireless channel.

- $\theta(\mu)$, defined as the decay rate of the probability $Pr\{D(\infty) \geq D_{max}\}$, corresponds to the Doppler spectrum. This can be seen from Figure 30. As shown in the figure, different Doppler rates give different $\theta(\mu)$. In addition, the figure shows that $\theta(\mu)$ increases with the Doppler rate. The reason for this is the following. As the Doppler rate increases, the degree of time diversity of the channel also increases. This implies lower delay-violation probability $Pr\{D(\infty) \geq D_{max}\}$, which leads to a larger decay rate $\theta(\mu)$. Therefore, $\theta(\mu)$ does reflect the Doppler spectrum of the underlying physical channel.

Note that a stationary Markov fading process (as is commonly assumed, for a physical wireless channel), $r(t)$, will always have a log-moment generating function $\Lambda(u)$. Therefore, the EC channel model is applicable to such a case. On the other hand, in Section 4.4.2, the simulation results show that the delay-violation probability $Pr\{D(\infty) \geq D_{max}\}$ does decrease exponentially with the delay bound D_{max} (see Figures 31 to 33). Therefore, our channel model is reasonable, not only from a theoretical viewpoint (*i.e.*, Markovian property of fading channels) but also from an experimental viewpoint (*i.e.*, the actual delay-violation probability decays exponentially with the delay bound).

4.3.4 QoS Guarantees Using Physical-layer Channel Models

In this section, we show that in contrast to our approach, which uses a link-layer model, the existing physical-layer channel models cannot be easily used to extract QoS guarantees. In Ref. [165], the authors attempt to model the wireless physical-layer channel using a discrete state representation. For example, Ref. [165] approximates a Rayleigh flat fading channel as a multi-state Markov chain, whose states are characterized by different bit error rates. With the multi-state Markov chain model, the performance of the link layer can be analyzed, but only at expense of enormous complexity. In this section, we outline a similar method for the Rayleigh flat fading channel at low SNRs.

As mentioned earlier, $\{\gamma(\mu), \theta(\mu)\}$ cannot be calculated using (27), in general, if only the marginal PDF at any time t and the Doppler spectrum are known. However, such an analytical calculation is possible for a Rayleigh flat fading channel in AWGN, albeit at very high complexity.

Suppose that the wireless channel is a Rayleigh flat fading channel in AWGN with Doppler spectrum $S(f)$. Assume that we have perfect causal knowledge of the channel gains. For example, the Doppler spectrum $S(f)$ from the Gans model [109] is the following

$$S(f) = \frac{1.5}{\pi f_m \sqrt{1 - (\frac{F}{f_m})^2}}, \tag{41}$$

where f_m is the maximum Doppler frequency; f_c is the carrier frequency; and $F = f - f_c$.

We show how to calculate the effective capacity for this channel. Denote a sequence of N_t measurements of the channel gain over the duration $[0, t]$, spaced at a time-interval δ apart, by $\mathbf{x} = [x(0), x(1), \cdots, x(N_t - 1)]$, where $x(n)$ $(n = 0, 1, \cdots, N_t - 1)$ are the complex-valued channel gains ($|x(n)|$ are therefore Rayleigh distributed). Without loss of generality, we have absorbed the constant noise variance into the definition of $x(n)$. The measurement $x(n)$ is a realization of a random variable sequence denoted by $X(n)$, which can be written as the vector $\mathbf{X} = [X(0), X(1), \cdots, X(N_t - 1)]$. The PDF of a random vector $\mathbf{X}$ for the Rayleigh fading channel is

$$f_{\mathbf{X}}(\mathbf{x}) = \frac{1}{\pi^{N_t} det(\mathbf{R})} e^{-\mathbf{x}^H \mathbf{R}^{-1} \mathbf{x}}, \tag{42}$$

where $\mathbf{R}$ is the covariance matrix of the random vector $\mathbf{X}$, $det(\mathbf{R})$ the determinant of matrix $\mathbf{R}$, and $\mathbf{x}^H$ the conjugate of $\mathbf{x}$. Now, to calculate the effective capacity, we first need to calculate,

$$E[e^{-u\tilde{S}(t)}] = E[e^{-u\int_0^t r(\tau)d\tau}]$$

$$\stackrel{(a)}{\approx} \int e^{-u(\sum_{n=0}^{N_t-1} \delta \times r(\tau_n))} f_{\mathbf{x}}(\mathbf{x})d\mathbf{x}$$

$$\stackrel{(b)}{=} \int e^{-u(\sum_{n=0}^{N_t-1} \delta \log(1+|x(n)|^2))} f_{\mathbf{x}}(\mathbf{x})d\mathbf{x}$$

$$\stackrel{(c)}{=} \int e^{-u(\sum_{n=0}^{N_t-1} \delta \log(1+|x(n)|^2))} \frac{1}{\pi^{N_t} det(\mathbf{R})} e^{-\mathbf{x}^H \mathbf{R}^{-1} \mathbf{x}} d\mathbf{x} \quad (43)$$

where (a) approximates the integral by a sum, (b) from the standard result on Gaussian channel capacity (*i.e.*, $r(\tau_n) = \log(1 + |x(n)|^2)$, where τ_n is the nth sampling epoch, and $|x(n)|$ is the modulus of $x(n)$), and (c) from Eq. (42). This gives the effective capacity (27) as,

$$\alpha(u) = \frac{-1}{u} \lim_{t\to\infty} \frac{1}{t} \log \int e^{-u(\sum_{n=0}^{N_t-1} \delta \log(1+|x(n)|^2))} \frac{1}{\pi^{N_t} det(\mathbf{R})} e^{-\mathbf{x}^H \mathbf{R}^{-1} \mathbf{x}} d\mathbf{x} \quad (44)$$

In general, the integral in (44) is of high dimension (*i.e.*, $2N_t$ dimensions) and it does not reduce to a simple form, except for the case of low SNR, where approximation can be made. Next, we show a simple form of (44), for the case of low SNR. We first simplify (43) as follows.

$$E[e^{-u\tilde{S}(t)}] \stackrel{(a)}{\approx} \int e^{-u\delta(\sum_{n=0}^{N_t-1} |x(n)|^2)} \frac{1}{\pi^{N_t} det(\mathbf{R})} e^{-\mathbf{x}^H \mathbf{R}^{-1} \mathbf{x}} d\mathbf{x}$$

$$\stackrel{(b)}{=} \int e^{-u\delta ||\mathbf{x}||^2} \frac{1}{\pi^{N_t} det(\mathbf{R})} e^{-\mathbf{x}^H \mathbf{R}^{-1} \mathbf{x}} d\mathbf{x}$$

$$\stackrel{(c)}{=} \frac{1}{\pi^{N_t} det(\mathbf{R})} \int e^{-\mathbf{x}^* (\mathbf{R}^{-1} + u\delta \mathbf{I}) \mathbf{x}} d\mathbf{x}$$

$$= \frac{1}{\pi^{N_t} det(\mathbf{R})} \times \pi^{N_t} det((\mathbf{R}^{-1} + u\delta \mathbf{I})^{-1})$$

$$= \frac{1}{det(u\delta \mathbf{R} + \mathbf{I})} \quad (45)$$

where (a) using the approximation $\log(1 + |x(n)|^2) \approx |x(n)|^2$ for Eq. (43) (if $|x(n)|$ is small, that is, low SNR), (b) by the definition of the norm of the vector $\mathbf{x}$, and (c) by the relation $||\mathbf{x}||^2 = \mathbf{x}^* \mathbf{x}$ (where $\mathbf{I}$ is identity matrix). We consider three cases of interest for Eq. (45):

- *Case 1 (special case):* Suppose $\mathbf{R} = r\mathbf{I}$, where $r = E|x(n)|^2$ is the average channel capacity. This case happens when a mobile moves very fast with respect to the sample period. From Eq. (45), we have

$$E[e^{-u\tilde{S}(t)}] \approx \frac{1}{det(u\delta\mathbf{R} + \mathbf{I})} = \frac{1}{(ur\delta + 1)^{N_t}} \stackrel{(a)}{=} \frac{1}{(ur \times \frac{t}{N_t} + 1)^{N_t}} \quad (46)$$

where (a) follows from the fact that the sample period δ is $\frac{t}{N_t}$.

As the number of samples $N_t \to \infty$, we have,

$$\lim_{N_t \to \infty} E[e^{-u\tilde{S}(t)}] \approx \lim_{N_t \to \infty} \frac{1}{(ur \times \frac{t}{N_t} + 1)^{N_t}} = e^{-urt} \quad (47)$$

Thus, in the limiting case, the Rayleigh fading channel reduces to an AWGN channel. Note that this result would not apply at high SNRs, because of the concavity of the $\log(\cdot)$ function. Since Case 1 has the highest degree of diversity, it is the best case for guaranteeing QoS, *i.e.*, it provides the largest effective capacity among all the Rayleigh fading processes with the same marginal PDF. It is also the best case for high SNR.

- *Case 2 (special case):* Suppose $\mathbf{R} = r\mathbf{1} \cdot \mathbf{1}^T$, where $(.)^T$ denotes matrix transpose, and $\mathbf{1} = [1, 1, \cdots, 1]^T$. Thus, all the samples are fully correlated, which could occur if the wireless terminal is immobile. From Eq. (45), we have,

$$E[e^{-u\tilde{S}(t)}] \approx \frac{1}{det(u\delta\mathbf{R} + \mathbf{I})} = \frac{1}{ur\delta N_t + 1} = \frac{1}{ur \times \frac{t}{N_t} \times N_t + 1} = \frac{1}{1 + urt}$$
$$(48)$$

Since Case 2 has the lowest degree of diversity, it is the worst case. Specifically, Case 2 provides zero effective capacity because a wireless terminal could be in a deep fade forever, making it impossible to guarantee any non-zero capacity. It is also the worst case for high SNR.

- *Case 3 (general case):* Denote the eigenvalues of matrix $\mathbf{R}$ by λ_n ($n = 0, 1, \cdots, N_t - 1$). Since $\mathbf{R}$ is symmetric, we have $\mathbf{R} = \mathbf{U}\mathbf{\Sigma}\mathbf{U}^H$, where $\mathbf{U}$ is a unitary matrix; $\mathbf{U}^H$ is its Hermitian; and the diagonal matrix

$\Sigma = diag(\lambda_0, \lambda_1, \cdots, \lambda_{N_t-1})$. From Eq. (45), we have,

$$
\begin{aligned}
E[e^{-u\tilde{S}(t)}] &\approx \frac{1}{det(u\delta\mathbf{R} + \mathbf{I})} \\
&= \frac{1}{det(u\delta\mathbf{U}\Sigma\mathbf{U}^H + \mathbf{U}\mathbf{U}^H)} \\
&= \frac{1}{det(\ \mathbf{U}\ diag(u\delta\lambda_0 + 1, u\delta\lambda_1 + 1, \cdots, u\delta\lambda_{N_t-1} + 1)\ \mathbf{U}^H\ )} \\
&= \frac{1}{\Pi_n(u\delta\lambda_n + 1)} \\
&= e^{-\sum_n \log(u\delta\lambda_n+1)}
\end{aligned}
\tag{49}
$$

Case 3 is the general case for a Rayleigh flat fading channel at low SNRs.

We now use the calculated $E[e^{-u\tilde{S}(t)}]$ to derive the log-moment generating function as,

$$
\begin{aligned}
\Lambda(-u) &= \lim_{t\to\infty} \frac{1}{t} \log E[e^{-u\tilde{S}(t)}] \\
&\overset{(a)}{\approx} \lim_{t\to\infty} \frac{1}{t} \log e^{-\sum_n \log(u\delta\lambda_n+1)} \\
&\overset{(b)}{=} \lim_{\Delta f\to 0} -\Delta f \sum_n \log(u\frac{\lambda_n}{B_w} + 1) \\
&\overset{(c)}{=} -\int \log(uS(f) + 1)df
\end{aligned}
\tag{50}
$$

where (a) follows from Eq. (49), (b) follows from the fact that the frequency interval $\Delta f = 1/t$ and the bandwidth $B_w = 1/\delta$, and (c) from the fact that the power spectral density $S(f) = \lambda_n/B_w$ and that the limit of a sum becomes an integral. This gives the effective capacity (27) as,

$$
\alpha(u) = \frac{\int \log(uS(f) + 1)df}{u}
\tag{51}
$$

Thus, the Doppler spectrum allows us to calculate $\alpha(u)$. The effective capacity function (51) can be used to guarantee QoS using Eq. (29).

Remark 4.1 We argue that even if we have *perfect* knowledge about the channel gains, it is hard to extract QoS metrics from the radio-layer channel model, in the general case. The effective capacity function (51) is valid only for a Rayleigh flat fading channel, *at low SNR*. At high SNR, the effective

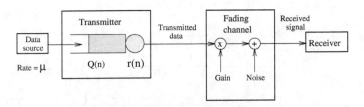

Figure 26: The queueing model used for simulations.

capacity for a Rayleigh fading channel is specified by the complicated integral in (44). To the best of our knowledge, a closed-form solution to (44) does not exist. It is clear that a numerical calculation of effective capacity is also very difficult, because the integral has a high dimension. Thus, it is difficult to extract QoS metrics from a radio-layer channel model, even for a Rayleigh flat fading channel. The extraction may not even be possible for more general fading channels. In contrast, the EC channel model that we have proposed can be easily translated into QoS metrics for a connection, and we have shown a simple estimation algorithm to estimate the EC model functions.

4.4 Simulation Results

In this section, we simulate a queueing system and demonstrate the performance of our algorithm for estimating the functions of the effective capacity link model. Section 4.4.1 describes the simulation setting, while Section 4.4.2 illustrates the performance of our estimation algorithm.

4.4.1 Simulation Setting

We simulate the discrete-time system depicted in Figure 26. In this system, the data source generates packets at a *constant* rate μ. Generated packets are first sent to the (infinite) buffer at the transmitter, whose queue length is $Q(n)$, where n refers to the n-th sample-interval. The head-of-line packet in the queue is transmitted over the fading channel at data rate $r(n)$. The fading channel has a random power gain $g(n)$. We use a fluid model, that is, the size of a packet is infinitesimal. In practical systems, the results presented here will have to be modified to account for finite packet sizes.

We assume that the transmitter has perfect knowledge of the current channel gains $g(n)$ at each sample-interval. Therefore, it can use rate-adaptive transmissions and ideal channel codes, to transmit packets without decoding errors. Thus, the transmission rate $r(n)$ is equal to the instantaneous (time-varying) capacity of the fading channel, as below,

$$r(n) = B_c \log_2(1 + g(n) \times P_0/\sigma_n^2) \tag{52}$$

where B_c denotes the channel bandwidth, and the transmission power P_0 and noise variance σ_n^2 are assumed to be constant.

The average SNR is fixed in each simulation run. We define r_{awgn} as the capacity of an equivalent AWGN channel, which has the same average SNR, i.e.,

$$r_{awgn} = B_c \log_2(1 + SNR_{avg}) \tag{53}$$

where $SNR_{avg} = E[g(n) \times P_0/\sigma^2] = P_0/\sigma^2$. We set $E[g(n)] = 1$. Then, we can eliminate B_c using Eqs. (52) and (53) as,

$$r(n) = \frac{r_{awgn} \log_2(1 + g(n) \times SNR_{avg})}{\log_2(1 + SNR_{avg})}. \tag{54}$$

In all the simulations, we set $r_{awgn} = 100$ kb/s.

In our simulations, the sample interval is set to 1 milli-second. This is not too far from reality, since 3G WCDMA systems already incorporate rate adaptation on the order of 10 milli-second [60].

Most simulation runs are 1000-second long; some simulation runs are 10000-second long in order to obtain good estimate of the actual delay-violation probability $Pr\{D(\infty) \geq D_{max}\}$ by the Monte Carlo method.

Denote $h(n)$ the voltage gain in the nth sample interval. We generate Rayleigh flat-fading voltage-gains $h(n)$ by a first-order auto-regressive (AR(1)) model as below. We first generate $\bar{h}(n)$ by

$$\bar{h}(n) = \kappa \times \bar{h}(n-1) + u_g(n), \tag{55}$$

where $u_g(n)$ are i.i.d. complex Gaussian variables with zero mean and unity variance per dimension. Then, we normalize $\bar{h}(n)$ and obtain $h(n)$ by

$$h(n) = \bar{h}(n) / \sqrt{\frac{2}{1 - \kappa^2}} = \bar{h}(n) \times \sqrt{\frac{1 - \kappa^2}{2}}. \tag{56}$$

It is clear that (56) results in $E[g(n)] = E[|h(n)|^2] = 1$. The coefficient κ determines the Doppler rate, i.e., the larger the κ, the smaller the Doppler rate. Specifically, the coefficient κ can be determined by the following procedure: 1) compute the coherence time T_c by [109, page 165]

$$T_c \approx \frac{9}{16\pi f_m}, \tag{57}$$

93

Table 4: Simulation parameters.

Channel	Maximum Doppler rate f_m	5 to 30 Hz
	AWGN channel capacity r_{awgn}	100 kb/s
	Average SNR	0/15 dB
	Sampling-interval T_s	1 ms
Source	Constant bit rate μ	30 to 85 kb/s

where the coherence time is defined as the time, over which the time auto-correlation function of the fading process is above 0.5; 2) compute the coefficient κ by[4]

$$\kappa = 0.5^{T_s/T_c}. \tag{58}$$

For Ricean fading, the voltage-gains $h(n)$ are generated by adding a constant to Rayleigh-fading voltage-gains (see [104] for detail).

Table 4 lists the parameters used in our simulations.

4.4.2 Performance of the Estimation Algorithm

We organize this section as follows. In Section 4.4.2.1, we estimate the functions $\{\gamma(\mu), \theta(\mu)\}$ of the effective capacity link model, from the measured x_n. Section 4.4.2.2 provides simulation results that demonstrate the relation between the physical channel and our link model. In Section 4.4.2.3, we show that the estimated EC functions accurately predict the QoS metric, under a variety of conditions.

4.4.2.1 Effective Capacity Model $\{\hat{\gamma}, \hat{\theta}\}$ Estimation

In the simulations, a Rayleigh flat fading channel is assumed. We simulate four cases: 1) $SNR_{avg} = 15$ dB and the maximum Doppler rate $f_m = 5$ Hz, 2) $SNR_{avg} = 15$ dB and $f_m = 30$ Hz, 3) $SNR_{avg} = 0$ dB and $f_m = 5$ Hz, and 4) $SNR_{avg} = 0$ dB and $f_m = 30$ Hz. Figures 27 and 28 show the estimated EC functions $\hat{\gamma}(\mu)$ and $\hat{\theta}(\mu)$. As the source rate μ increases from 30 kb/s to 85 kb/s, $\hat{\gamma}(\mu)$ increases, indicating a higher buffer occupancy, while $\hat{\theta}(\mu)$ decreases, indicating a slower decay of the delay-violation probability. Thus, the delay-violation probability is expected to increase, with increasing source rate μ. From Figure 27, we also observe that SNR has a substantial impact

---

[4]The auto-correlation function of the AR(1) process is κ^m, where m is the number of sample intervals. Solving $\kappa^{T_c/T_s} = 0.5$ for κ, we obtain (58).

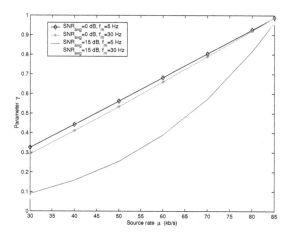

Figure 27: Estimated function $\hat{\gamma}(\mu)$ vs. source rate μ.

on $\hat{\gamma}(\mu)$. This is because higher SNR results in larger channel capacity, which leads to smaller probability that a packet will be buffered, *i.e.*, smaller $\hat{\gamma}(\mu)$. In contrast, Figure 27 shows that f_m has little effect on $\hat{\gamma}(\mu)$. The reason is that $\hat{\gamma}(\mu)$ reflects the marginal CDF of the underlying fading process, rather than the Doppler spectrum.

4.4.2.2 Physical Interpretation of Link Model $\{\hat{\gamma}, \hat{\theta}\}$

To illustrate that different physical channel induces different parameters $\{\hat{\gamma}, \hat{\theta}\}$, we simulate two kinds of channels, *i.e.*, a Rayleigh flat fading channel and a Rician flat fading channel. For the Rayleigh channel, we set the average SNR to 15 dB. For the Rician channel, we set the K factor[5] to 3 dB. We simulate two scenarios: A) changing the source rate while fixing the Doppler rate at 30 Hz, and B) changing the Doppler rate while fixing the source rate, *i.e.*, $\mu = 85$ kb/s.

The result for scenario A is shown in Figure 29. For comparison, we also plot the marginal CDF (*i.e.*, Rayleigh/Rician CDF) of the physical channel in the same figure. The marginal CDF for Rayleigh channel, *i.e.*, the probability

---

[5]The K factor is defined as the ratio between the deterministic signal power A^2 and the variance of the multipath $2\sigma_m^2$, *i.e.*, $K = A^2/(2\sigma_m^2)$.

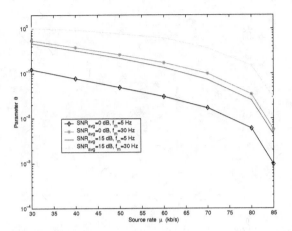

Figure 28: Estimated function $\hat{\theta}(\mu)$ vs. source rate μ.

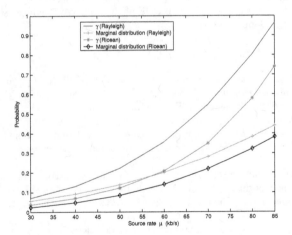

Figure 29: Marginal CDF and $\gamma(\mu)$ vs. source rate μ.

that the SNR falls below a threshold SNR_{th}, is

$$Pr\{SNR \leq SNR_{th}\} = 1 - e^{-SNR_{th}/SNR_{avg}} \tag{59}$$

Similar to (54), we have the source rate

$$\mu = \frac{r_{awgn} \log_2(1 + SNR_{th})}{\log_2(1 + SNR_{avg})} \tag{60}$$

Solving (60) for SNR_{th}, we obtain

$$SNR_{th} = (1 + SNR_{avg})^{\frac{\mu}{r_{awgn}}} - 1 \tag{61}$$

Using (59) and (61), we plot the marginal CDF of the Rayleigh channel, as a function of source rate μ. Similarly, we plot the marginal CDF of the Ricean channel, as a function of source rate μ.

As shown in Figure 29, different marginal CDF at the physical layer yields different $\hat{\gamma}(\mu)$ at the link layer. We observe that $\hat{\gamma}(\mu)$ and marginal CDF have similar behavior, *i.e.*, 1) both increases with the source rate μ; 2) if one channel has a larger outage probability than another channel, it also has a larger $\hat{\gamma}(\mu)$ than the other channel. For example, in Figure 29, the Rayleigh channel has a larger outage probability and a larger $\hat{\gamma}(\mu)$ than the Ricean channel. Thus, the probability of non-empty buffer, $\hat{\gamma}(\mu)$, is similar to marginal CDF, *i.e.*, outage probability.

Figures 30 and 31 show the result for scenario B. From Figure 30, it can be seen that different Doppler rate at the physical layer leads to different $\hat{\theta}(\mu)$ at the link layer. In addition, the figure shows that $\hat{\theta}(\mu)$ increases with the Doppler rate. This is reasonable since the increase of the Doppler rate leads to the increase of time diversity, resulting in a larger decay rate $\hat{\theta}(\mu)$ of the delay-violation probability. Therefore, $\hat{\theta}(\mu)$ corresponds to the Doppler spectrum of the physical channel.

Figure 31 shows the actual delay-violation probability $Pr\{D(\infty) \geq D_{max}\}$ vs. the delay bound D_{max}, for various Doppler rates. It can be seen that the actual delay-violation probability decreases exponentially with the delay bound D_{max}, for all the cases. This justifies the use of an exponential bound (40) in predicting QoS, thereby justifying our link model $\{\hat{\gamma}, \hat{\theta}\}$.

4.4.2.3 *Accuracy of the QoS Metric Predicted by $\hat{\gamma}$ and $\hat{\theta}$*

In the previous section, the simulation results have justified the use of $\{\hat{\gamma}, \hat{\theta}\}$ in predicting QoS. In this section, we evaluate the accuracy of such a prediction.

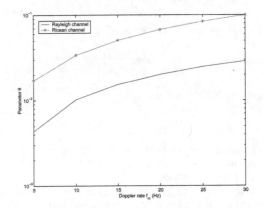

Figure 30: θ vs. Doppler rate f_m.

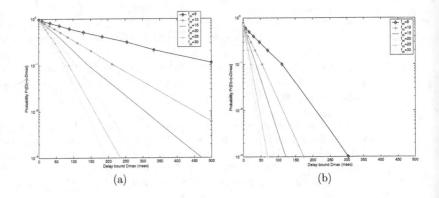

Figure 31: Actual delay-violation probability vs. D_{max}, for various Doppler rates: (a) Rayleigh fading and (b) Ricean fading.

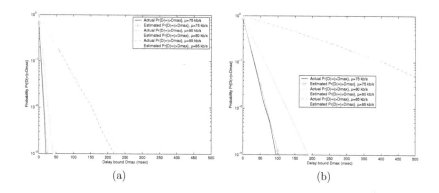

(a)   (b)

Figure 32: Prediction of delay-violation probability, when the average SNR is
(a) 15 dB and (b) 0 dB.

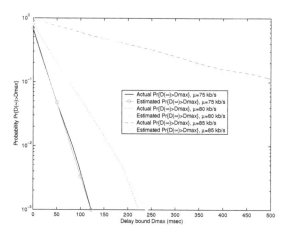

Figure 33: Prediction of delay-violation probability, when $f_m = 5$ Hz.

To test the accuracy, we use $\hat{\gamma}$ and $\hat{\theta}$ to calculate the delay-bound violation probability $Pr\{D(\infty) \geq D_{max}\}$ (using (40)), and then compare the estimated probability with the actual (*i.e., measured*) $Pr\{D(\infty) \geq D_{max}\}$.

To show the accuracy, we simulate three scenarios. In the first scenario, the source rates μ are 75/80/85 kb/s, which loads the system as light/moderate/heavy, respectively. For all three cases, we simulate a Rayleigh flat fading channel with $SNR_{avg} = 15$ dB, $r_{awgn} = 100$ kb/s and $f_m = 30$ Hz. Figure 32(a) plots the actual and the estimated delay-bound violation probability $Pr\{D(\infty) >> D_{max}\}$ as a function of D_{max}. As predicted by (40), the delay-violation probability follows an exponential decrease with D_{max}. Furthermore, the estimated $Pr\{D(\infty) \geq D_{max}\}$ is close to the actual $Pr\{D(\infty) \geq D_{max}\}$.

In the second scenario, we also set $r_{awgn} = 100$ kb/s and $f_m = 30$ Hz, but change the average SNR to 0 dB. Figure 32(b) shows that the conclusions drawn from the first scenario still hold. Thus, our estimation algorithm gives consistent performance over different SNRs also.

In the third scenario, we set $SNR_{avg} = 15$ dB and $r_{awgn} = 100$ kb/s, but we change the Doppler rate f_m to 5 Hz. Figure 33 shows that the conclusions drawn from the first scenario still hold. Thus, our estimation algorithm consistently predicts the QoS metric under different Doppler rate f_m.

In summary, the simulations illustrate that our EC link model, together with the estimation algorithm, predict the actual QoS accurately.

4.5   Modeling for Frequency Selective Fading Channels

In Section 4.3, we modeled the wireless channel in terms of two 'effective capacity' functions; namely, the probability of non-empty buffer $\gamma(\mu)$ and the QoS exponent $\theta(\mu)$. Furthermore, we developed a simple and efficient algorithm to estimate the EC functions $\{\gamma(\mu), \theta(\mu)\}$. For flat-fading channels (*e.g.*, Rayleigh fading with a specified Doppler spectrum), the simulation results have shown that the actual QoS metric is closely approximated by the QoS metric predicted by the EC channel model and its estimation algorithm, under various scenarios. On the other hand, for frequency selective fading channels with high degrees of frequency diversity, a refinement is useful to characterize $\gamma(\mu)$ further. The technique we use is large deviations theory for many sources.

In this section, we use the duality between the distribution of a queue with superposition of N i.i.d. sources, and the distribution of a queue with a frequency-selective fading channel that consists of N i.i.d. sub-channels, to propose a channel model, specified by three functions $\{\beta^{(c)}(\mu), \eta^{(c)}(\mu), \theta_1(\mu)\}$.

The remainder of this section is organized as follows. Section 4.5.1 presents large deviation results for a queue with many inputs, which provide us the method to model frequency-selective fading channels. In Section 4.5.2, we describe a modified effective capacity model for frequency-selective fading channels. Section 4.5.3 shows simulation results that demonstrate the accuracy of the proposed channel model.

4.5.1 Large Deviation Results for a Queue with Many Inputs

In [18], Botvich and Duffield obtained a large deviation result for the queue at a multiplexer with N inputs. To state their results, let $Q^N(\infty)$ be the length of the steady-state queue due to a superposition of N i.i.d. stationary sources, served at constant rate $N \times r$ (r fixed). Denote by $A^N(t)$ ($t \geq 0$) the amount of aggregate traffic from the N sources over the time interval $[0,t)$. Assume that the many-source-asymptotic[6] log-moment generating function of $A^N(t)$, defined as

$$\Lambda_t(u) = \lim_{N \to \infty} \frac{1}{Nt} \log E[e^{u(A^N(t) - Nrt)}], \qquad (62)$$

exists for all $u > 0$ and $t > 0$.

Define Λ_t^* the Legendre-Fenchel transform of Λ_t, through

$$\Lambda_t^*(B) = \sup_u (B \times u - \Lambda_t(u)). \qquad (63)$$

Under appropriate conditions, paralleling (23) for large buffer asymptotic, Botvich and Duffield's result for many source asymptotic [18] is

$$Pr\{Q^N(\infty) \geq B\} \sim e^{-N \times I(B/N)} \qquad \text{as } N \to \infty, \qquad (64)$$

where $I(B) = \inf_{t>0} t\Lambda_t^*(B/t)$, and $f(x) \sim g(x)$ means that $\lim_{x \to \infty} f(x)/g(x) = 1$. Eq. (64) yields the approximation

$$Pr\{Q^N(\infty) \geq B\} \approx e^{-N \times I(B/N)}, \qquad (65)$$

By introducing a prefactor $\beta^{(s)}(r)$, a more accurate approximation was proposed in [28] as below,

$$Pr\{Q^N(\infty) \geq B\} \approx \beta^{(s)}(r) \times e^{-N \times I(B/N)} \qquad (66)$$

$$\approx \beta^{(s)}(r) \times e^{-\eta^{(s)}(r) \times N} \times e^{-\theta_B(N,r) \times B}, \qquad (67)$$

---

[6]Many source asymptotic is for the case where the number of sources N goes to ∞ while large buffer asymptotic is for the case where the buffer size B goes to ∞.

where $\beta^{(s)}(r)$ and $\eta^{(s)}(r)$ are functions of r, $\theta_B(N,r)$ is a function of N and r, and $\eta^{(s)}(r)$ and $\theta_B(N,r)$ satisfy [18]

$$\lim_{B \to \infty} (I(B) - \theta_B(N,r) \times B) = \eta^{(s)}(r), \qquad (68)$$

and

$$\eta^{(s)}(r) = - \lim_{t \to \infty} t \Lambda_t(\theta_B(N,r)), \qquad (69)$$

under appropriate conditions. The quantity $\beta^{(s)}(r) \times e^{-\eta^{(s)}(r) \times N}$ can be regarded as the probability that the buffer is non-empty; this probability decays exponentially as the number of sources N increases, which is different from $\gamma(r)$ in Section 4.3.1.

If the quantity of interest is the steady-state delay $D^N(\infty)$, then the probability of $D^N(\infty)$ exceeding a delay bound D_{max} satisfies

$$Pr\{D^N(\infty) \geq D_{max}\} \approx \beta^{(s)}(r) \times e^{-\eta^{(s)}(r) \times N} \times e^{-\theta^{(s)}(N,r) \times D_{max}}, \qquad (70)$$

where $\theta^{(s)}(N,r) = \theta_B(N,r) \times N \times r$. Thus, the triplet $\{\beta^{(s)}(r), \eta^{(s)}(r), \theta^{(s)}(N,r)\}$ models the aggregate source.

In the following section, we use the duality between traffic modeling ($\{\beta^{(s)}(r), \eta^{(s)}(r), \theta^{(s)}$ and channel modeling to propose a link-layer model for frequency selective fading channels, specified by a triplet $\{\beta^{(c)}(\mu), \eta^{(c)}(\mu), \theta_N(\mu)\}$. It is clear that we intend $\{\beta^{(c)}(\mu), \eta^{(c)}(\mu), \theta_N(\mu)\}$ to be the channel duals of the source functions $\{\beta^{(s)}(r), \eta^{(s)}(r), \theta^{(s)}(N,r)\}$. Just as the constant *channel rate* r is used in source traffic modeling, we use the constant *source traffic rate* μ in modeling the channel.

4.5.2   Channel Modeling

Consider a queue of infinite buffer size supplied by a data source of *constant* data rate μ, served by $1/N$ fraction of a frequency-selective fading channel that consists of N i.i.d. sub-channels. The queue is served by $1/N$ fraction of the channel to keep the system load[7] constant as N increases. The large deviation results in Section 4.5.1 can be easily adapted to this case. The difference is that whereas in Section 4.5.1, the source rate was variable while the channel capacity per source was constant, in this section, the source rate is constant while the channel capacity is variable. Similar to (70), it can be shown that the probability of $D(\infty)$ exceeding a delay bound D_{max} satisfies

---

[7]The system load is defined as the ratio of the expected source rate to the ergodic channel capacity.

$$Pr\{D(\infty) \geq D_{max}\} \approx \gamma_N(\mu) \times e^{-\theta_N(\mu) \times D_{max}} \tag{71}$$

where the functions $\{\gamma_N(\mu), \theta_N(\mu)\}$ characterize the frequency-diversity channel with N independent sub-channels, and the function $\gamma_N(\mu)$ can be approximated by

$$\gamma_N(\mu) \approx \beta^{(c)}(\mu) \times e^{-\eta^{(c)}(\mu) \times N} \tag{72}$$

Assuming equality in (72), we can easily derive a method to estimate $\eta^{(c)}(\mu)$ and $\beta^{(c)}(\mu)$ as below

$$\eta^{(c)}(\mu) = -\log(\gamma_N(\mu)/\gamma_1(\mu))/(N-1), \tag{73}$$

and

$$\beta^{(c)}(\mu) = \gamma_1(\mu) \times e^{\eta^{(c)}(\mu)}. \tag{74}$$

where $\gamma_1(\mu)$ and $\gamma_N(\mu)$ can be estimated by Eq. (36).

For the case where the sub-channels are i.i.d., a simplification occurs. Let $r_N(t)$ be the instantaneous channel capacity of $1/N$ fraction of a frequency-selective fading channel with N i.i.d. sub-channels, at time t. Then, the *effective capacity function* of $r_N(t)$ is defined as

$$\alpha_N(u) = -\lim_{t\to\infty} \frac{1}{ut} \log E[e^{-u \int_0^t r_N(\tau)d\tau}], \ \forall \ u > 0, \tag{75}$$

if it exists. Let $r_1(t)$ be the instantaneous channel capacity of one sub-channel of the frequency-selective fading channel, at time t. Then, the *effective capacity function* of $r_1(t)$ is defined as

$$\alpha_1(u) = -\lim_{t\to\infty} \frac{1}{ut} \log E[e^{-u \int_0^t r_1(\tau)d\tau}], \ \forall \ u > 0. \tag{76}$$

According to Eq. (30), the QoS exponents $\theta_N(\mu)$ and $\theta_1(\mu)$ are defined as

$$\theta_N(\mu) = \mu \alpha_N^{-1}(\mu), \tag{77}$$

and

$$\theta_1(\mu) = \mu \alpha_1^{-1}(\mu), \tag{78}$$

respectively, and the two QoS exponents have a relation specified by Proposition 4.1.

Proposition 4.1 *The QoS exponents $\theta_N(\mu)$ and $\theta_1(\mu)$ satisfy*

$$\theta_N(\mu) = N \times \theta_1(\mu). \tag{79}$$

103

For a proof of Proposition 4.1, see Appendix A.2.

From (71), (72), and (79), we have

$$Pr\{D(\infty) \geq D_{max}\} \approx \beta^{(c)}(\mu) \times e^{-\eta^{(c)}(\mu) \times N} \times e^{-N \times \theta_1(\mu) \times D_{max}}. \quad (80)$$

So, the functions $\{\beta^{(c)}(\mu), \eta^{(c)}(\mu), \theta_1(\mu)\}$ sufficiently characterize the QoS $Pr\{D(\infty) \geq D_{max}\}$ for a frequency-selective fading channel, consisting of *arbitrary* N i.i.d. sub-channels. There is no need to directly estimate $\gamma_N(\mu)$ and $\theta_N(\mu)$ for arbitrary N, using Eq. (36) through (39).

The EC channel model for frequency-selective fading channels and its application are summarized below.

1. $\{\beta^{(c)}(\mu), \eta^{(c)}(\mu), \theta_1(\mu)\}$ is the EC channel model.

2. $\{\beta^{(c)}(\mu), \eta^{(c)}(\mu), \theta_1(\mu)\}$ can be estimated by (74), (73), and (36) through (39), respectively.

3. Given the EC channel model, the QoS $\{\mu, D_{max}, \varepsilon\}$ can be computed by Eq. (80), where $\varepsilon = Pr\{D(\infty) \geq D_{max}\}$.

4.5.3 Simulation Results

The simulation settings are the same as that in Section 4.4.1 except that the flat fading channel is replaced by a frequency selective fading channel. That is, the channel voltage gain $h_i(n)$ of each sub-channel i $(i = 1, 2, \cdots, N)$ in a frequency selective fading channel, is assumed to be Rayleigh-distributed and is generated by an AR(1) model as below. We first generate $\bar{h}_i(n)$ by

$$\bar{h}_i(n) = \kappa \times \bar{h}_i(n - 1) + v_i(n), \quad (81)$$

where $v_i(n)$ are i.i.d. complex Gaussian variables with zero mean and unity variance per dimension. Then, we normalize $\bar{h}_i(n)$ and obtain $h_i(n)$ by

$$h_i(n) = \bar{h}_i(n) \times \sqrt{\frac{1 - \kappa^2}{2}}. \quad (82)$$

We simulate the discrete-time system depicted in Figure 26. In this system, the data source generates packets at a *constant* rate μ. Generated packets are first sent to the (infinite) buffer at the transmitter, whose queue length is $Q(n)$, where n refers to the n-th sample-interval. The head-of-line packet in the queue is transmitted over the fading channel at data rate $r(n)$. Since

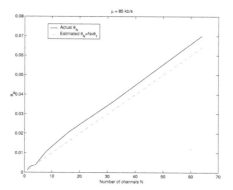

Figure 34: Actual and estimated θ_N vs. N.

the transmission rate $r(n)$ is equal to $1/N$ of the sum of the instantaneous capacities of the N sub-channels, we have

$$r(n) = \frac{1}{N}\sum_{i=1}^{N} r_i(n) = \frac{\sum_{i=1}^{N} r_{awgn}\log_2(1 + |h_i(n)|^2 \times SNR_{avg})}{N\log_2(1 + SNR_{avg})}. \quad (83)$$

In all the simulations in this section, we fix the following parameters: $r_{awgn} = 100$ kb/s, $\kappa = 0.98$, and $SNR_{avg} = 0$ dB. The sample interval T_s is set to 1 milli-second and each simulation run is 1000-second long in all scenarios.

In the next section, we use simulation results to show the accuracy of the channel model $\{\beta^{(c)}(\mu), \eta^{(c)}(\mu), \theta_1(\mu)\}$.

4.5.3.1 Accuracy of Channel Model $\{\beta^{(c)}(\mu), \eta^{(c)}(\mu), \theta_1(\mu)\}$

In the simulations, we first directly estimate γ_N and θ_N for various number of sub-channels N, using Eqs. (36) through (39); then we estimate $\{\beta^{(c)}(\mu), \eta^{(c)}(\mu), \theta_1(\mu)\}$ by (74), (73), and (36) through (39), respectively.

Figure 34 shows the actual and estimated θ_N vs. N for $\mu = 85$ kb/s. The actual θ_N is meant to be the θ_N directly measured by Eqs. (36) through (39) for the case with N sub-channels; the estimated θ_N is meant to be $N \times \theta_1$, $i.e.$, Eq. (79), where θ_1 is measured by Eqs. (36) through (39) for the case with one sub-channel. The figure indicates that 1) the actual θ_N linearly increase with N, justifying the linear relation in (79), and 2) the estimated θ_N can serve as a rough estimate of the actual θ_N.

Figure 35 shows the actual and estimated γ_N vs. N for various source rate μ. The actual γ_N is meant to be the γ_N directly measured by Eqs. (36)

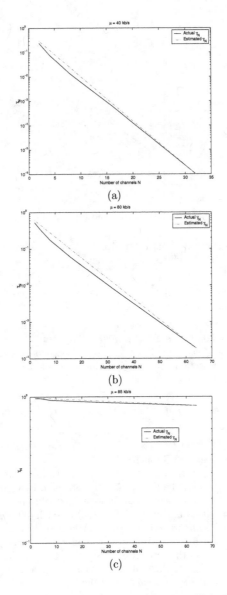

Figure 35: Actual and estimated γ_N vs. N: (a) $\mu = 40$ kb/s, (b) $\mu = 60$ kb/s, and (c) $\mu = 85$ kb/s.

106

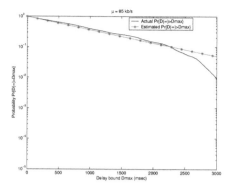

Figure 36: Actual and estimated delay-bound violation probability for $\mu = 85$ kb/s and $N = 1$.

through (39) for the case with N sub-channels; the estimated γ_N is obtained by Eq. (72), where $\eta^{(c)}(\mu)$ and $\beta^{(c)}(\mu)$ are estimated by (73) and (74), respectively. The figure demonstrates that 1) the actual γ_N decrease exponentially with N, justifying the exponential relation in (72), and 2) the estimated γ_N is close to the actual γ_N.

Figures 36 and 37 show the actual and estimated delay-bound violation probability $Pr\{D(\infty) \geq D_{max}\}$ vs. the delay bound D_{max}, for various N and $\mu = 85$ kb/s. The actual $Pr\{D(\infty) \geq D_{max}\}$ is obtained by directly measuring the queue; the estimated $Pr\{D(\infty) \geq D_{max}\}$ is obtained by Eq. (80), where $\{\beta^{(c)}(\mu), \eta^{(c)}(\mu), \theta_1(\mu)\}$ are estimated by (74), (73), and (36) through (39), respectively. The figures illustrate that the estimated $Pr\{D(\infty) \geq D_{max}\}$ agrees with the actual $Pr\{D(\infty) \geq D_{max}\}$. It is clear that as the number of sub-channels N increases, $Pr\{D(\infty) \geq D_{max}\}$ decreases for a fixed D_{max}. This indicates that frequency diversity improves the delay performance of a wireless channel. We will show how to utilize frequency diversity to improve delay performance in Chapter 6.

Figure 38 shows the actual and estimated $Pr\{D(\infty) \geq D_{max}\}$ vs. the delay bound D_{max}, for various μ and $N = 4$. The actual and estimated $Pr\{D(\infty) \geq D_{max}\}$ are obtained by the same ways as Figure 37. Figure 38 indicates that the estimated $Pr\{D(\infty) \geq D_{max}\}$ gives good agreement with the actual $Pr\{D(\infty) \geq D_{max}\}$ for various data rates μ.

Figure 39 shows the actual and estimated $Pr\{D(\infty) \geq D_{max}\}$ vs. the delay bound D_{max}, for various μ and $N = 4$, with a simplified estimation method. The actual $Pr\{D(\infty) \geq D_{max}\}$ is obtained by the same way as Figure 37.

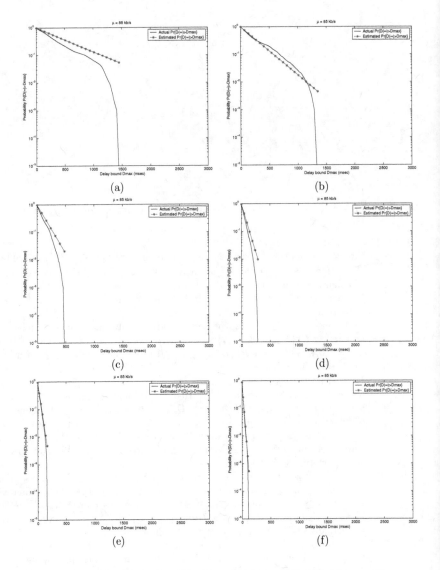

Figure 37: Actual and estimated delay-bound violation probability for $\mu = 85$ kb/s and various N: (a) $N = 2$, (b) $N = 4$, (c) $N = 8$, (d) $N = 16$, (e) $N = 32$, and (f) $N = 64$.

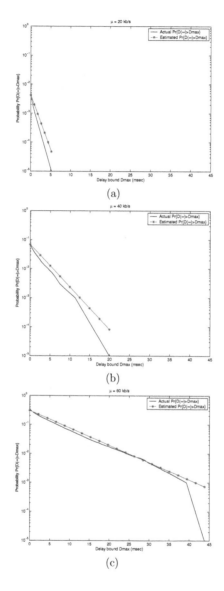

Figure 38: Actual and estimated delay-bound violation probability for $N = 4$ channels: (a) $\mu = 20$ kb/s, (b) $\mu = 40$ kb/s, and (c) $\mu = 60$ kb/s.

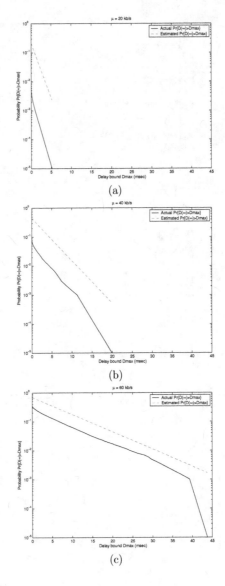

Figure 39: Actual and simply estimated delay-bound violation probability for $N = 4$ channels: (a) $\mu = 20$ kb/s, (b) $\mu = 40$ kb/s, and (c) $\mu = 60$ kb/s.

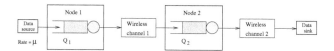

Figure 40: A network with tandem wireless links.

Different from Eq. (80), the estimated $Pr\{D(\infty) \geq D_{max}\}$ is obtained by a simplified estimate as below,

$$Pr\{D(\infty) \geq D_{max}\} = \gamma_1(\mu) \times e^{-N \times \theta_1(\mu) \times D_{max}}, \qquad (84)$$

where $\gamma_1(\mu)$ and $\theta_1(\mu)$ are estimated by Eqs. (36) through (39) for the case with one sub-channel. Compared with Figure 38, Figure 39 indicates that if $\gamma_N(\mu)$ in (71) is replaced by $\gamma_1(\mu)$, the estimated $Pr\{D(\infty) \geq D_{max}\}$ would be very conservative. Hence, the estimation of $\gamma_N(\mu)$ is necessary.

In summary, by estimating functions $\{\beta^{(c)}(\mu), \eta^{(c)}(\mu), \theta_1(\mu)\}$ and using Eq. (80), we can obtain the QoS $Pr\{D(\infty) \geq D_{max}\}$ for a frequency-selective fading channel, consisting of arbitrary N i.i.d. sub-channels, with reasonable accuracy.

4.6 Effective Capacity-based QoS Measures

The following property is needed in the propositions in this section.

Property 4.1 *(i) The asymptotic log-moment generation function $\Lambda(u)$ defined in (26) is finite for all $u \in \mathbb{R}$. (ii) $\Lambda(u)$ is differentiable for all $u \in \mathbb{R}$.*

4.6.1 QoS Measures for Wireless Networks

In this section, we consider two basic network structures for wireless networks: one with only tandem wireless links (see Figure 40) and the other with only parallel wireless links (see Figure 41). In the following, Propositions 4.2 and 4.3 give QoS measures for these two network structures, respectively.

Denote $r_k(t)$ $(k = 1, \cdots, K)$ the instantaneous capacity of channel k at time t. For a network with K tandem links, define the service $\tilde{S}(t_0, t)$, for $t \geq 0$ and any $t_0 \in [0, t]$, by

$$\tilde{S}(t_0, t) = \inf_{t_0 \leq t_1 \leq \cdots \leq t_{K-1} \leq t_K = t} \left\{ \sum_{k=1}^{K} \int_{t_{k-1}}^{t_k} r_k(\tau) \mathrm{d}\tau \right\}, \qquad (85)$$

and the asymptotic log-moment generating function

$$\Lambda_{tandem}(-u) = \lim_{t \to \infty} \frac{1}{t} \log E[e^{-u\tilde{S}(0,t)}] \qquad (86)$$

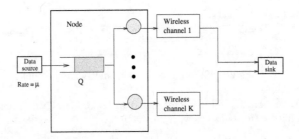

Figure 41: A network with parallel wireless links.

where $\tilde{S}(0,t)$ is defined by (85); also define the effective capacity of channel k by

$$\alpha_k(u) = -\lim_{t\to\infty} \frac{1}{ut} \log E[e^{-u\int_0^t r_k(\tau)\mathrm{d}\tau}], \ \forall \ u > 0. \qquad (87)$$

Proposition 4.2 *Assume that the log-moment generating function $\Lambda_{tandem}(u)$ defined by (86) satisfies Property 4.1. Given the effective capacity functions $\{\alpha_k(u), k = 1, \cdots, K\}$ of K tandem links and an external arrival process with constant rate μ, the end-to-end delay $D(\infty)$ experienced by the traffic traversing the K tandem links satisfies*

$$\limsup_{D_{max}\to\infty} \frac{1}{D_{max}} \log Pr\{D(\infty) > D_{max}\} \leq -\theta, \qquad if \ \alpha(\theta/\mu) > \mu, \qquad (88)$$

and

$$\lim_{D_{max}\to\infty} \frac{1}{D_{max}} \log Pr\{D(\infty) > D_{max}\} = -\theta^*, \qquad where \ \alpha(\theta^*/\mu) = \mu, \quad (89)$$

where $\alpha(u) = -\Lambda_{tandem}(-u)/u$. Moreover, the effective capacity $\alpha(u)$ satisfies

$$\alpha(u) \leq \min_k \alpha_k(u). \qquad (90)$$

For a proof of Proposition 4.2, see Appendix A.3. Note that the capacity processes of the tandem channels are not required to be independent in Proposition 4.2.

Proposition 4.3 *Assume that the log-moment generating function $\Lambda_k(u)$ of each channel k in the network satisfies Property 4.1. Given the effective capacity functions $\{\alpha_k(u), k = 1, \cdots, K\}$ of K independent parallel links and*

112

an external arrival process with constant rate μ, the end-to-end delay $D(\infty)$ experienced by the traffic traversing the K parallel links satisfies

$$\limsup_{D_{max} \to \infty} \frac{1}{D_{max}} \log Pr\{D(\infty) > D_{max}\} \le -\theta, \qquad if \; \alpha(\theta/\mu) > \mu, \qquad (91)$$

and

$$\lim_{D_{max} \to \infty} \frac{1}{D_{max}} \log Pr\{D(\infty) > D_{max}\} = -\theta^*, \qquad where \; \alpha(\theta^*/\mu) = \mu, \quad (92)$$

where $\alpha(u) = \sum_{k=1}^{K} \alpha_k(u)$.

For a proof of Proposition 4.3, see Appendix A.4.

Propositions 4.2 and 4.3 suggest the following approximation

$$Pr\{D(\infty) > D_{max}\} \approx e^{-\theta^* \times D_{max}}, \qquad (93)$$

for large D_{max}. In addition, $\alpha(u)$ specified in Propositions 4.2 and 4.3 can be regarded as the effective capacity of the equivalent channel of the network, which consists of tandem links only or independent parallel links only. In Sections 4.6.2 to 4.6.4, we will use $\alpha(u)$ to characterize the equivalent channel of the network; and we will use (92) only since (92) is tighter than (91).

4.6.2 QoS Measures for Variable-Bit-Rate Sources

In this section, we develop QoS measures for the case where the sources generate traffic at variable bit-rates (VBR). We consider two classes of VBR sources: leaky-bucket constrained arrival (see Section 2.2) and exponential process with its effective bandwidth function known (see Section 2.2.1). Propositions 4.4 and 4.5 provide QoS measures for these two classes of VBR sources, respectively.

Proposition 4.4 *Assume that a wireless network consists of tandem links only or independent parallel links only; the effective capacity function of the equivalent channel of the wireless network is characterized by $\alpha(u)$; and the log-moment generating function $\Lambda_k(u)$ of each channel k in the network satisfies Property 4.1. Given an external arrival process constrained by a leaky bucket with bucket size $\sigma^{(s)}$ and token generating rate $\lambda_s^{(s)}$, the end-to-end delay $D(\infty)$ experienced by the traffic traversing the network satisfies*

$$\lim_{D_{max} \to \infty} \frac{1}{D_{max} - \sigma^{(s)}/\lambda_s^{(s)}} \log Pr\{D(\infty) > D_{max}\} = -\theta^*, \; where \; \alpha(\theta^*/\lambda_s^{(s)}) = \lambda_s^{(s)}.$$

$$(94)$$

113

For a proof of Proposition 4.4, see Appendix A.5. Eq. (94) suggests the following approximation

$$Pr\{D(\infty) > D_{max}\} \approx e^{-\theta^* \times (D_{max} - \sigma^{(s)}/\lambda_s^{(s)})}, \qquad (95)$$

for large D_{max}.

Proposition 4.5 *Assume that a wireless network consists of tandem links only or independent parallel links only; the effective capacity function of the equivalent channel of the wireless network is characterized by $\alpha(u)$; an external arrival process is characterized by its effective bandwidth function $\alpha^{(s)}(u)$; and the log-moment generating function $\Lambda_k(u)$ of each channel k in the network and the log-moment generating function $\Lambda^{(s)}(u)$ of the external arrival process satisfy Property 4.1. Denote u^* the unique solution of the following equation*

$$\alpha^{(s)}(u) = \alpha(u). \qquad (96)$$

The end-to-end delay $D(\infty)$ experienced by the traffic traversing the network satisfies

$$\lim_{D_{max} \to \infty} \frac{1}{D_{max}} \log Pr\{D(\infty) > D_{max}\} = -\theta^*, \ \ where \ \theta^* = u^* \times \alpha^{(s)}(u^*). \quad (97)$$

For a proof of Proposition 4.5, see Appendix A.6.

Note that a single-link network is a special case in Propositions 4.4 and 4.5.

4.6.3 QoS Measures for Packetized Traffic

In previous sections, we assumed fluid traffic. In this section, we extend the QoS measures obtained previously for the fluid model to the case with packetized traffic. This is important since in practical situations, the packet size is not negligible (not infinitesimal as in fluid model).

We assume the propagation delay of a wireless link is negligible, and the service at a network node is *non-cut-through*, *i.e.*, no packet is eligible for service until its last bit has arrived. We also assume a wireless network consists of tandem links only or parallel links only. For a network with tandem links only, the number of hops in the network is determined by the number of tandem links in the network; for a network with parallel links only, the number of hops in the network is one. We consider two cases: 1) a constant-bit-rate source with constant packet size, and 2) a variable-bit-rate source with variable packet size. Propositions 4.6 and 4.7 give QoS measures for these two cases, respectively.

114

Proposition 4.6 *Assume that a wireless network consists of tandem links only or independent parallel links only; the effective capacity function of the equivalent channel of the wireless network is characterized by $\alpha(u)$; the log-moment generating function $\Lambda_k(u)$ of each channel k in the network satisfies Property 4.1; and the network consists of N hops. Given an external arrival process with constant bit rate μ and constant packet size L_c, the end-to-end delay $D(\infty)$ experienced by the traffic traversing the network satisfies*

$$\lim_{D_{max} \to \infty} \frac{1}{D_{max} - N \times L_c/\mu} \log Pr\{D(\infty) > D_{max}\} = -\theta^*, \text{ where } \alpha(\theta^*/\mu) = \mu.$$
(98)

For a proof of Proposition 4.6, see Appendix A.7. Eq. (98) suggests the following approximation

$$Pr\{D(\infty) > D_{max}\} \approx e^{-\theta^*(D_{max} - N \times L_c/\mu)},$$
(99)

for large D_{max}.

Proposition 4.7 *Assume that a wireless network consists of tandem links only or independent parallel links only; the effective capacity function of the equivalent channel of the wireless network is characterized by $\alpha(u)$; the log-moment generating function $\Lambda_k(u)$ of each channel k in the network satisfies Property 4.1; and the network consists of N hops. Given a traffic flow having maximum packet size L_{max} and constrained by a leaky bucket with bucket size $\sigma^{(s)}$ and token generating rate $\lambda_s^{(s)}$, the end-to-end delay $D(\infty)$ experienced by the traffic traversing the network satisfies*

$$\lim_{D_{max} \to \infty} \frac{1}{D_{max} - N \times L_{max}/\lambda_s^{(s)} - \sigma^{(s)}/\lambda_s^{(s)}} \log Pr\{D(\infty) > D_{max}\} = -\theta^*,$$
(100)

where $\alpha(\theta^*/\lambda_s^{(s)}) = \lambda_s^{(s)}$.

For a proof of Proposition 4.7, see Appendix A.8. Eq. (98) suggests the following approximation

$$Pr\{D(\infty) > D_{max}\} \approx e^{-\theta^*(D_{max} - N \times L_{max}/\lambda_s^{(s)} - \sigma^{(s)}/\lambda_s^{(s)})},$$
(101)

for large D_{max}.

Note that a single-link network is a special case in Propositions 4.6 and 4.7.

4.6.4 QoS Measures for Wireless Channels with Non-negligible Propagation Delay

In previous sections, we assumed the propagation delay of a wireless link is negligible. In this section, we extend the QoS measures obtained previously to the situation where the propagation delay of a wireless link is not negligible. We consider two cases: 1) a fluid source with a constant rate, and 2) a variable-bit-rate source with variable packet size. Propositions 4.8 and 4.9 give QoS measures for these two cases, respectively.

Proposition 4.8 *Assume that a wireless network consists of tandem links only or independent parallel links only; the effective capacity function of the equivalent channel of the wireless network is characterized by $\alpha(u)$; the log-moment generating function $\Lambda_k(u)$ of each channel k in the network satisfies Property 4.1; the network consists of N hops; and the i-th hop $(i = 1, \cdots, N)$ incurs a constant propagation delay d_i. Given a fluid traffic flow with constant rate μ, the end-to-end delay $D(\infty)$ experienced by the traffic traversing the network satisfies*

$$\lim_{D_{max} \to \infty} \frac{1}{D_{max} - \sum_{i=1}^{N} d_i} \log Pr\{D(\infty) > D_{max}\} = -\theta^*, \text{ where } \alpha(\theta^*/\mu) = \mu. \tag{102}$$

For a proof of Proposition 4.8, see Appendix A.9. Eq. (102) suggests the following approximation

$$Pr\{D(\infty) > D_{max}\} \approx e^{-\theta^*(D_{max} - \sum_{i=1}^{N} d_i)}, \tag{103}$$

for large D_{max}.

Proposition 4.9 *Assume that a wireless network consists of tandem links only or independent parallel links only; the effective capacity function of the equivalent channel of the wireless network is characterized by $\alpha(u)$; the log-moment generating function $\Lambda_k(u)$ of each channel k in the network satisfies Property 4.1; the network consists of N hops; and the i-th hop $(i = 1, \cdots, N)$ incurs a constant propagation delay d_i. Given a traffic flow having maximum packet size L_{max} and constrained by a leaky bucket with bucket size $\sigma^{(s)}$ and token generating rate $\lambda_s^{(s)}$, the end-to-end delay $D(\infty)$ experienced by the traffic traversing the network satisfies*

$$\lim_{D_{max} \to \infty} \frac{1}{D_{max} - N \times L_{max}/\lambda_s^{(s)} - \sigma^{(s)}/\lambda_s^{(s)} - \sum_{i=1}^{N} d_i} \log Pr\{D(\infty) > D_{max}\} = -\theta^*, \tag{104}$$

where $\alpha(\theta^*/\lambda_s^{(s)}) = \lambda_s^{(s)}$.

For a proof of Proposition 4.9, see Appendix A.10. Eq. (104) suggests the following approximation

$$Pr\{D(\infty) > D_{max}\} \approx e^{-\theta^*(D_{max} - N \times L_{max}/\lambda_s^{(s)} - \sigma^{(s)}/\lambda_s^{(s)} - \sum_{i=1}^{N} d_i)}, \qquad (105)$$

for large D_{max}.

4.7 Summary

Efficient bandwidth allocation and QoS provisioning over wireless links, demand a simple and effective wireless channel model. In this chapter, we modeled a wireless channel from the perspective of the communication *link-layer*. This is in contrast to existing channel models, which characterize the wireless channel at the *physical-layer*. Specifically, we modeled the wireless link in terms of two 'effective capacity' functions; namely, the probability of nonempty buffer $\gamma(\mu)$ and the QoS exponent $\theta(\mu)$. The QoS exponent is the inverse of a function which we call *effective capacity* (EC). The EC channel model is the dual of the effective bandwidth source traffic model, used in wired networks. Furthermore, we developed a simple and efficient algorithm to estimate the EC functions $\{\gamma(\mu), \theta(\mu)\}$. Simulation results show that the actual QoS metric is closely approximated, by the QoS metric predicted by the EC channel model and its estimation algorithm, under various scenarios.

We have provided key insights about the relations between the EC channel model and the physical-layer channel, *i.e.*, $\gamma(\mu)$ corresponds to the marginal CDF (*e.g.*, Rayleigh/Ricean distribution) while $\theta(\mu)$ is related to the Doppler spectrum. The EC channel model has been justified not only from a theoretical viewpoint (*i.e.*, Markov property of fading channels) but also from an experimental viewpoint (*i.e.*, the delay-violation probability does decay exponentially with the delay).

In this chapter, we have obtained link-layer QoS measures for various scenarios: flat-fading channels, frequency-selective fading channels, multi-link wireless networks, variable-bit-rate sources, packetized traffic, and wireless channels with non-negligible propagation delay. The QoS measures obtained can be easily translated into traffic envelope and service curve characterizations, which are popular in wired networks, such as ATM and IP, to provide guaranteed services. Therefore, we believe that the EC channel model, which was specifically constructed keeping in mind these link-layer QoS measures, will find wide applicability in future wireless networks that need QoS provisioning.

In summary, our EC channel model has the following features: simplicity of implementation, efficiency in admission control, and flexibility in allocating

117

bandwidth and delay for connections. In addition, our channel model provides a general framework, under which physical-layer fading channels such as AWGN, Rayleigh fading, and Ricean fading channels can be studied.

Armed with the new channel model, we investigate its use in designing admission control, resource reservation, and scheduling algorithms, for efficient support of QoS guarantees, in Chapters 5 and 6.

CHAPTER 5

QOS PROVISIONING: A SINGLE SHARED CHANNEL

5.1 Introduction

Providing quality of service (QoS), such as delay and rate guarantees, is an important objective in the design of future packet cellular networks [60]. However, this requirement poses a challenge in wireless network design, because wireless channels have low reliability, and time varying signal strength, which may cause severe QoS violations. Further, the capacity of a wireless channel is severely limited, making efficient bandwidth utilization a priority.

An effective way to increase the outage capacity [13] of a time-varying channel is the use of diversity. The idea of diversity is to create multiple *independent* signal paths between the transmitter and the receiver so that higher outage capacity can be obtained. Diversity can be achieved over time, space, and frequency. These traditional diversity methods are essentially applicable to a single-user link. Recently, however, Knopp and Humblet [72] introduced another kind of diversity, which is inherent in a wireless network with multiple users sharing a time-varying channel. This diversity, termed *multiuser diversity* [52], comes from the fact that different users usually have *independent* channel gains for the same shared medium. With multiuser diversity, the strategy of maximizing the total Shannon (ergodic) capacity is to allow at any time slot only the user with the best channel to transmit. This strategy is called Knopp and Humblet's (K&H) scheduling. Results [72] have shown that the K&H scheduling can increase the total (ergodic) capacity dramatically, in the absence of delay constraints, as compared to the traditionally used (weighted) round robin (RR) scheduling where each user is *a priori* allocated fixed time slots.

The K&H scheduling intends to maximize ergodic capacity, which pertains to situations of infinite tolerable delay. However, under this scheme, a user in a fade of an arbitrarily long period will not be allowed to transmit during this period, resulting in an arbitrarily long delay; therefore, this scheme provides no delay guarantees and thus is not suitable for delay-sensitive applications, such as voice or video. To mitigate this problem, Bettesh and Shamai

119

[9] proposed an algorithm, which strikes a balance between throughput and delay constraints. This algorithm combines the K&H scheduling with an RR scheduling, and it can achieve lower delay than the K&H scheduling while obtaining a capacity gain over a pure RR scheduling. However, it is very complex to theoretically relate the QoS obtained by this algorithm to the control parameters of the algorithm, and thus cannot be used to guarantee a specified QoS. Furthermore, a direct (Monte Carlo) measurement of QoS achieved, using the queueing behavior resulting from the algorithm, requires an excessively large number of samples, so that it becomes practically infeasible.

Another typical approach is to use dynamic programming [10] to design a scheduler that can increase capacity, while also maintaining QoS guarantees. But this approach suffers from the curse of dimensionality, since the size of the dynamic program state space grows exponentially with the number of users and with the delay requirement.

To address these problems, this chapter proposes an approach, which simplifies the task of explicit provisioning of QoS guarantees while achieving efficiency in utilizing wireless channel resources. Specifically, we design our scheduler based on the K&H scheduling, but shift the burden of QoS provisioning to the resource allocation mechanism, thus simplifying the design of the scheduler. Such a partitioning would be meaningless if the resource allocation problem now becomes complicated. However, we are able to solve the resource allocation problem efficiently using the method of *effective capacity* developed in Chapter 4. Effective capacity captures the effect of channel fading on the queueing behavior of the link, using a computationally simple yet accurate model, and thus, is the critical device we need to design an efficient resource allocation mechanism.

Our results show that compared to the RR scheduling, our approach can substantially increase the statistical delay-constrained capacity (defined later) of a fading channel, when delay requirements are not very tight. For example, in the case of low signal-to-noise-ratio (SNR) and ergodic Rayleigh fading, our scheme can achieve approximately $\sum_{k=1}^{K} \frac{1}{k}$ gain for K users with loose-delay requirements, as expected from [72]. But more importantly, when the delay bound is not loose, so that simple-minded K&H scheduling does not directly apply, our scheme can achieve a capacity gain, and yet meet the QoS requirements.

The remainder of this chapter is organized as follows. In Section 5.2, we discuss multiuser diversity, which is our key technique to increase capacity. Section 5.3 presents efficient QoS provisioning mechanisms and shows how to use multiuser diversity and effective capacity to achieve a performance gain while yet satisfying QoS constraints. In Section 5.4, we present the simulation results that demonstrate the performance gain of our scheme. In Section 5.5,

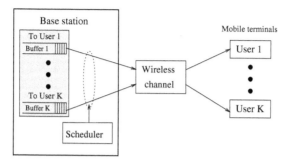

Figure 42: Downlink scheduling architecture for multiple users sharing a wireless channel.

we summarize this chapter.

5.2 Multiuser Diversity with QoS Constraints

We organize this section as below. In Section 5.2.1, we discuss the technique of multiuser diversity. Section 5.2.2 describes statistical QoS and its relation with effective capacity. In Section 5.2.3, we illustrate the limitation of traditional QoS provisioning methods through an example.

5.2.1 Multiuser Diversity

We first describe the model. Figure 42 shows the architecture for scheduling multiuser traffic over a fading (time-varying) time-slotted wireless channel. A cellular wireless network is assumed, and the downlink is considered, where a base station transmits data to K mobile user terminals, each of which requires certain QoS guarantees. The channel fading processes of the users are assumed to be stationary, ergodic and independent of each other. A single cell is considered, and interference from other cells is modeled as background noise with constant variance. In the base station, packets destined to different users are put into separate queues. We assume a block fading channel model [13], which assumes that user channel gains are constant over a time duration of length T_s (T_s is assumed to be small enough that the channel gains are constant, yet large enough that ideal channel codes can achieve capacity over that duration). Therefore, we partition time into 'frames' (indexed as $t = 0, 1, 2, \ldots$), each of length T_s. Thus, each user k has a time-varying channel power gain $g_k(t), k = 1, \ldots, K$, which varies with the frame index t; and $g_k(t) = |h_k(t)|^2$, where $h_k(t)$ is the voltage gain of the channel for the kth user. The base station is assumed to know the current and past values of $g_k(t)$. The capacity of the channel for the kth user, $c_k(t)$, is

$$c_k(t) = \log_2(1 + g_k(t) \times P_0/\sigma_n^2) \qquad \text{bits/symbol} \qquad (106)$$

where the maximum transmission power P_0 and noise variance σ_n^2 are assumed to be constant and equal for all users. We divide each frame of length T_s into infinitesimal time slots, and assume that the channel can be shared by several users, in the same frame. Further, we assume a *fluid model* for packet transmission, where the base station can allot *variable fractions* of a channel frame to a user, over time. The system described above could be, for example, an idealized time-division multiple access (TDMA) system, where the frame of each channel consists of TDMA time slots which are infinitesimal. Note that in a practical TDMA system, there would be a finite number of finite-length time slots in each frame.

To provide QoS guarantees, we propose an architecture, which consists of scheduling, admission control, and resource allocation (presented in Section 5.3). Since the channel fading processes of the users are assumed to be independent of each other, we can potentially utilize multiuser diversity to increase capacity, as mentioned in Section 5.1. Thus, *to maximize the ergodic capacity* (*i.e.*, in the absence of delay constraints), the (optimal) K&H schedule at any time instant t, is to transmit the data of the user with the largest gain $g_k(t)$ [72]. The ergodic channel capacity achieved by such a K&H scheduler is $c_{max} = \mathbf{E}[\max[c_1(t), c_2(t), \cdots, c_K(t)]]$. The ergodic channel capacity gain of the K&H scheduler over a RR scheduler is $c_{max}/\mathbf{E}[c_1(t)]$. The following proposition specifies the ergodic channel capacity gain achieved by the K&H scheduler.

Proposition 5.1 *Assume that the K users in the system have channel gains, which are i.i.d. in user k ($k = 1, \cdots, K$) and are stationary in time t. For Rayleigh fading channels (i.e., having exponentially-distributed channel power gains), at low SNR, we have the approximation, $c_{max}/\mathbf{E}[c_1(t)] \approx \sum_{k=1}^{K} \frac{1}{k} \approx \log(K+1)$ for large K.*

For a proof of Proposition 5.1, see Appendix A.11. At high SNR, the ergodic channel capacity gain is smaller.

Notice that K&H scheduling can result in a user experiencing an arbitrarily long duration of outage, because of its failure to obtain the channel. Thus, it becomes important to efficiently compute the QoS obtained by the user, in a K&H scheduled system. A direct approach may be to model each $g_k(t)$ as a Markov process, and analyze the Markov process resulting from the

122

K&H scheduler. It is apparent that this direct approach is computationally intractable, since the large state space of the joint Markov process of all the users would need to be analyzed and the complexity of this queueing analysis is exponential in the number of users. We will show an example to illustrate this point in Section 5.2.3. In essence, the main contribution of this chapter is to show that we can compute the QoS obtained by the user, in a K&H scheduled system, efficiently and accurately, using the concept of effective capacity.

5.2.2 Statistical QoS and Effective Capacity

We first formally define statistical QoS, which characterizes the user requirement. First, consider a single-user system, where the user is allotted a single time varying channel (thus, there is no scheduling involved). Assume that the user source has a fixed rate r_s and a specified delay bound D_{max}, and requires that the delay-bound violation probability is not greater than a certain value ε, that is,

$$Pr\{D(\infty) > D_{max}\} \leq \varepsilon, \qquad (107)$$

where $D(\infty)$ is the steady-state delay experienced by a flow, and $Pr\{D(\infty) > D_{max}\}$ is the probability of $D(\infty)$ exceeding a delay bound D_{max}. Then, we say that the user is specified by the (statistical) QoS triplet $\{r_s, D_{max}, \varepsilon\}$. Even for this simple case, it is not immediately obvious as to which QoS triplets are feasible, for the given channel, since a rather complex queueing system (with an arbitrary channel capacity process) will need to be analyzed. The key contribution of Chapter 4 was to introduce a concept of statistical delay-constrained capacity termed *effective capacity*, which allows us to obtain a simple and efficient test, to check the feasibility of QoS triplets for a single time-varying channel. Chapter 4 did not deal with scheduling and the channel processes resulting from it.

In this chapter, we show that the effective capacity concept can be applied to the K&H scheduled channel, and is precisely the critical device that we need to solve the QoS constrained multiuser diversity problem. Next, we describe the relation between statistical QoS and effective capacity.

Let $r(t)$ be the instantaneous channel capacity at time t. The *effective capacity function* $\alpha(u)$ of $r(t)$ is defined by (28). Consider a queue of infinite buffer size supplied by a data source of *constant* data rate μ. From Chapter 4, we know that if $\alpha(u)$ exists, then the probability of $D(\infty)$ exceeding a delay bound D_{max} satisfies

$$Pr\{D(\infty) > D_{max}\} \approx e^{-\theta(\mu)D_{max}}, \qquad (108)$$

where the QoS exponent function $\theta(\mu)$ is related to $\alpha(u)$ via (30), and $\theta(\mu)$ depends only on the channel capacity process $r(t)$. Note that (108) is a simplified version of (108), where $\gamma(\mu)$ is negligible for large D_{max}. Once $\theta(\mu)$ has been measured for a given channel, it can be used to check the feasibility of QoS triplets. Specifically, a QoS triplet $\{r_s, D_{max}, \varepsilon\}$ is feasible if $\theta(r_s) \geq \rho$, where $\rho \doteq -\log \varepsilon / D_{max}$. Thus, we can use the effective capacity model $\alpha(u)$ (or equivalently, the function $\theta(\mu)$) to relate the channel capacity process $r(t)$ to statistical QoS. Since our effective capacity method predicts an exponential dependence (108) between ε and D_{max}, we can henceforth consider the QoS pair $\{r_s, \rho\}$ to be equivalent to the QoS triplet $\{r_s, D_{max}, \varepsilon\}$, with the understanding that $\rho = -\log \varepsilon / D_{max}$.

In Section 4.3.2, we presented a simple and efficient algorithm to estimate $\theta(\mu)$ by direct measurement on the queueing behavior resulting from $r(t)$. In Section 5.4.2.1, we show that the estimation algorithm converges quickly, as compared with directly measuring the QoS.

5.2.3 Limitation of Traditional QoS Provisioning

As mentioned in Section 4.1, to explicitly enforce QoS guarantees in a system such as the K&H scheduler, a typical procedure involves four steps: 1) channel measurement, 2) channel modeling, 3) deriving QoS measures, and 4) relating the control parameters of QoS provisioning mechanisms to the derived QoS measures. However, the traditional QoS provisioning approach uses physical-layer channel models, which incurs *high complexity* in characterizing the relation between the control parameters and the calculated QoS measures. Next, we show an example to illustrate this.

Example 5.1 *Our objective here is to use the K&H scheduler to provide explicit QoS guarantees. The setting is the same as that in Section 5.2.1, i.e., one wireless channel is shared by K users, which require identical QoS specified by a QoS triplet $\{r_s, D_{max}, \varepsilon\}$; the users have independent channel gains; the K&H scheduler at the base station knows the current and past channel gains perfectly; the K&H scheduler allots the channel to the user with largest channel gain.*

The problem is: given the QoS requirement $\{r_s, D_{max}, \varepsilon\}$, what is the minimum fraction of a channel frame that can satisfy this QoS requirement? Denote β ($\beta \in (0, 1]$) the fraction of a channel frame allocated to the users. The solution by the traditional approach is the following:

1. *Model each users channel capacity $c_k(t)$ by a discrete-time Markov chain $X_k(t)$, where k is the index of a user.*

2. *Model the capacity assigned by the K&H scheduler to user 1. We denote this capacity by $\tilde{c}_1(t)$. The capacity $\tilde{c}_1(t)$ is equal to $\beta \times c_1(t)$ if user 1 has*

the largest channel power gain; otherwise, $\tilde{c}_1(t)$ is equal to zero. That is,

$$\tilde{c}_1(t) = \begin{cases} \beta \times c_1(t) & \text{if } X_1(t) = \max_{k=1,\cdots,K} X_k(t) \\ 0 & \text{otherwise} \end{cases} \qquad (109)$$

From (108), it is clear that $\tilde{c}_1(t)$ depends on the channel gains of all the K users and hence $\tilde{c}_1(t)$ is characterized by a Markov chain with K-dimensional state, which is specified by $\{X_1(t), \cdots, X_K(t)\}$.

3. *Derive the QoS measure $Pr\{D(\infty) > D_{max}\}$. From queueing theory, we know that if the service is a Markov chain with K-dimensional state, then obtaining the distribution of the queue has an exponential complexity, that is, in the order of M^K, where M is the number of states in $X_k(t)$. Once we derive the distribution of the queue, we can obtain $Pr\{D(\infty) > D_{max}\}$.*

4. *Find the minimum β such that the resulting delay-bound violation probability $Pr\{D(\infty) > D_{max}\} \le \varepsilon$. If no feasible $\beta \in (0,1]$ is found, which means there is not enough resource to support the requested QoS, then reject the connection request. Otherwise, the minimum feasible $\beta_{min} = \min \beta$ is allocated for the K&H scheduler to use. Here, β_{min} is the control parameter of the K&H scheduler.*

Note that the queueing analysis does not result in a closed-form relation between the control parameter β_{min} and the QoS requirement $\{r_s, D_{max}, \varepsilon\}$. One may suggest to 1) solve the queueing problem off-line (having an exponential complexity), 2) store the relation $\{r_s, D_{max}, \varepsilon; \mathbf{P}_K; \beta_{min}\}$ in a table where $\mathbf{P}_K$ is the K-dimensional transition probability matrix for $\{X_1(t), \cdots, X_K(t)\}$, and 3) look up the table when a connection request arrives. However, the size of the table is exponential in the number of users K since the number of all possible $\mathbf{P}_K$ is in the order of M^K.

Example 5.1 shows that the traditional approach incurs a complexity that is exponential in the number of users, to determine what percentage of the channel resource should be allocated to the K&H scheduler, so that a specified QoS can be satisfied. The reason of the high complexity is that the traditional approach employs physical-layer channel models, which use high-dimensional Markov chains. To address this problem, in the next section, we describe how to apply the effective capacity technique to designing simple and efficient QoS provisioning in a K&H scheduled system.

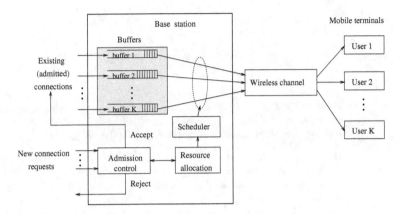

Figure 43: QoS provisioning architecture in a base station.

5.3 QoS Provisioning with Multiuser Diversity

The key problem is, how to utilize multiuser diversity while yet satisfying the individual QoS constraints of the K users. To cope with this problem, we design a QoS provisioning architecture, which utilizes multiuser diversity and effective capacity.

We assume the same setting as in Section 5.2.1. Figure 43 shows our QoS provisioning architecture in a base station, consisting of three components, namely, admission control, resource allocation, and scheduling. When a new connection request comes, we first use a resource allocation algorithm to compute how much resource is needed to support the requested QoS. Then the admission control module checks whether the required resource can be satisfied. If yes, the connection request is accepted; otherwise, the connection request is rejected. For admitted connections, packets that belong to different connections[1] are put into separate queues. The scheduler decides, in each frame t, how to schedule packets for transmission, based on the *current* channel gains $g_k(t)$ and the amount of resource allocated.

In the following sections, we describe our schemes for scheduling, admission control and resource allocation in detail. In Section 5.3.1, we consider the homogeneous case, in which all users have the same QoS requirements $\{r_s, D_{max}, \varepsilon\}$ or equivalently the same QoS pair $\{r_s, \rho = -\log \varepsilon / D_{max}\}$ and

---

[1] We assume that each mobile user is associated with only one connection.

also the same channel statistics (*e.g.*, similar Doppler rates), so that all users need to be assigned equal channel resources. Section 5.3.2 addresses the heterogeneous case, in which users have different QoS pairs $\{r_s, \rho\}$ and/or different channel statistics.

5.3.1 Homogeneous Case

5.3.1.1 Scheduling

As explained in Section 5.1, we simplify the scheduler, by shifting the burden of guaranteeing user QoS to the resource allocation module. Therefore, our scheduler is a simple combination of the K&H and the RR scheduling.

Section 5.2 explained that in any frame t, the K&H scheduler transmits the data of the user with the largest gain $g_k(t)$. However, the QoS of a user may be satisfied by using only a fraction of the frame $\beta \leq 1$. Therefore, it is the function of the resource allocation algorithm to allot the minimum required β to the user. This will be described in Section 5.3.1.2. It is clear that the K&H scheduling attempts to utilize multiuser diversity to maximize the throughput.

On the other hand, the RR scheduler allots to every user k, a fraction $\zeta \leq 1/K$ of *each* frame, where ζ again needs to be determined by the resource allocation algorithm. Thus the RR scheduling attempts to provide tight QoS guarantees, at the expense of decreased throughput, in contrast to the K&H scheduling.

Our scheduler is a joint K&H/RR scheme, which attempts to maximize the throughput, while yet providing QoS guarantees. In each frame t, its operation is the following. First, find the user $k^*(t)$ such that it has the largest channel gain among all users. Then, schedule user $k^*(t)$ with $\beta + \zeta$ fraction of the frame; schedule each of the other users $k \neq k^*(t)$ with ζ fraction of the frame. Thus, a fraction β of the frame is used by the K&H scheduling, while simultaneously, a total fraction $K\zeta$ of the frame is used by the RR scheduling. The total usage of the frame is $\beta + K\zeta \leq 1$.

5.3.1.2 Admission Control and Resource Allocation

The scheduler described in Section 5.3.1.1 is simple, but it needs the frame fractions $\{\zeta, \beta\}$ to be computed and reserved. This function is performed at the admission control and resource allocation phase.

Since Section 5.3.1 addresses the homogeneous case with K users, without loss of generality, denote $\alpha_{K,\zeta,\beta}(u)$ the effective capacity function of user $k = 1$ under the joint K&H/RR scheduling (henceforth called 'joint scheduling'), with frame shares ζ and β respectively, *i.e.*, denote the capacity process allotted to user 1 by the joint scheduler as the process $r(t)$ and then compute $\alpha_{K,\zeta,\beta}(u)$

127

using (28). The corresponding QoS exponent function $\theta_{K,\varsigma,\beta}(\mu)$ can be found via (30). Note that $\theta_{K,\varsigma,\beta}(\mu)$ is a function of number of users K. Then, the admission control and resource allocation scheme for users requiring the QoS pair $\{r_s, \rho\}$ is as below,

$$\underset{\{\varsigma,\beta\}}{\text{minimize}} \quad K\varsigma + \beta \tag{110}$$

$$\text{subject to} \quad \theta_{K,\varsigma,\beta}(r_s) \geq \rho, \tag{111}$$

$$K\varsigma + \beta \leq 1, \tag{112}$$

$$\varsigma \geq 0, \quad \beta \geq 0 \tag{113}$$

The minimization in (110) is to minimize the total frame fraction used. (111) ensures that the QoS pair $\{r_s, \rho\}$ of each user is feasible. Furthermore, Eqs. (111)–(113) also serve as an admission control test, to check availability of resources to serve this set of users. Actually, we only need to measure the $\theta_{K,\varsigma,\beta}(\cdot)$ functions for different ratios of ς/β due to Proposition 5.2.

Proposition 5.2 *For $\lambda > 0$, the following equation holds*

$$\theta_{K,\varsigma,\beta}(\mu) = \theta_{K,\lambda\varsigma,\lambda\beta}(\lambda\mu). \tag{114}$$

For a proof of Proposition 5.2, see Appendix A.12.

To summarize, given the fading channel and QoS of K homogeneous users, we use the following procedure to achieve multiuser diversity gain with QoS provisioning:

1. Estimate $\theta_{K,\varsigma,\beta}(\mu)$, directly from the queueing behavior, for various values of $\{\varsigma, \beta\}$.

2. Determine the optimal $\{\varsigma, \beta\}$ pair that satisfies users' QoS, while minimizing frame usage.

3. If admission control is passed, provide the joint scheduler with the optimal ς and β, for simultaneous RR and K&H scheduling, respectively.

This summary indicates that our approach needs to address the following issues. Chapter 4 showed the usefulness of the effective capacity concept, only for a single-user system. But, it is not obvious that the $\theta_{K,\varsigma,\beta}(\mu)$ estimate will converge quickly in the multiuser scenario, or even that $\theta_{K,\varsigma,\beta}(\mu)$ can accurately predict QoS via (108) (although, theoretically, the prediction is accurate asymptotically for large D_{max}). Further, it needs to be seen whether the QoS can be controlled by $\{\varsigma, \beta\}$. Last, we also need to show that our scheme can provide a substantial capacity gain, over the RR scheduling. These issues will be addressed via simulations in Section 5.4.

5.3.1.3 Improvement

The aforementioned admission control and resource allocation, *i.e.*, Eqs. (110)–(113), may not be efficient in terms of resource usage, when K is large and ρ takes a medium value. The reason is that for a large number of users K, the K&H scheduling causes large delays and thus the joint K&H/RR will reduce to the RR only. Partitioning the K users into groups, each group scheduled separately using the K&H/RR, can improve channel utilization. In the following, we show a modified K&H/RR that accomplishes this effect.

Before describing the scheme, we need to introduce some notations. Suppose that the K users are partitioned into M groups (obviously, $1 \leq M \leq K$). Each group m ($m = 1, 2, \cdots, M$) has K_m users with channel characterization (*i.e.*, QoS exponent function) $\theta_{K_m, \zeta_m, \beta_m}(\mu)$, where $\{\zeta_m, \beta_m\}$ are the frame shares assigned to group m, for the joint K&H/RR scheduling. Obviously, $\sum_{m=1}^{M} K_m = K$.

Next, we describe scheduling and resource allocation/admission control, respectively.

Scheduling

For *each group* m, we use the joint K&H/RR scheduler with frame shares $\{\zeta_m, \beta_m\}$. In each frame t, it works as follows. First, find the user $k_m^*(t)$ such that it has the largest channel gain among K_m users of group m (not among K users). Then, schedule user $k_m^*(t)$ with $\beta_m + \zeta_m$ fraction of the frame; schedule each of other group-m users $k \neq k_m^*(t)$ with ζ_m fraction of the frame. Thus, the total usage of the frame by all M group is $\sum_{m=1}^{M}(K_m \zeta_m + \beta_m) \leq 1$.

Admission Control and Resource Allocation

The scheduler described above requires the frame fractions $\{\zeta_m, \beta_m\}$ to be computed and reserved. This function is performed at the admission control and resource allocation phase. With dividing users into M groups, the admission control and resource allocation scheme for K users requiring the QoS pair $\{r_s, \rho\}$ is as below,

$$\underset{\{M, K_m, \zeta_m, \beta_m\}}{\text{minimize}} \quad \sum_{m=1}^{M}(K_m \zeta_m + \beta_m) \tag{115}$$

$$\text{subject to} \quad \theta_{K_m, \zeta_m, \beta_m}(r_s) \geq \rho, \qquad \forall m \tag{116}$$

$$\sum_{m=1}^{M}(K_m \zeta_m + \beta_m) \leq 1, \tag{117}$$

$$\zeta_m \geq 0, \qquad \beta_m \geq 0, \qquad \forall m, \tag{118}$$

$$M \in \{1, 2, \cdots, K\}, \qquad \sum_{m=1}^{M} K_m = K \tag{119}$$

Eq. (115) is to minimize the total frame fraction used. (116) ensures that the QoS pair $\{r_s, \rho\}$ of each user is feasible. Furthermore, Eqs. (116)–(119) also serve as an admission control test, to check availability of resources to serve the K users. If $M = 1$, Eqs. (115)–(119) reduced to Eqs. (110)–(113). Therefore, the optimal solution in Eqs. (110)–(113) is a feasible solution of Eqs. (115)–(119). As a result, the resource allocation/admission control in Eqs. (115)–(119) is at least as efficient as that in Eqs. (110)–(113), if not more efficient.

Note that the improvement on efficiency achieved in Eqs. (115)–(119) is at the cost of complexity.

5.3.2 Heterogeneous Case

For the heterogeneous case, in which users have different QoS pairs $\{r_s, \rho\}$ and/or different channel statistics, the admission control/resource allocation problem can also be formulated, similar to Eqs. (115)–(119), as minimizing the resource usage over M, partitioning of the K users, ζ_m and β_m. But solving this minimization problem has an exponential complexity, *i.e.*, $O(K^K)$, since we have to try all the possible combinations. To reduce the complexity, we design a sub-optimal algorithm, which has a complexity of $O(K \log K)$. We consider the following two cases.

Case 1: the K users have different channel statistics but the same $\{r_s, \rho\}$.

Figure 44 shows the flow chart of our algorithm for the resource allocation. The basic operations of this algorithm are sorting and partitioning. Sorting the users is to facilitate partitioning the users. Partitioning is achieved through tests, which recursively check whether adding a user to a K&H scheduled group can reduce the channel usage.

Next we describe the algorithm. According to Figure 44, we first measure the function $\mu_k(\theta = \rho)$ for each user k, where $\mu_k(\theta)$ is the inverse function of $\theta(\mu)$ defined in (30); note that there is no scheduling involved in (30). Then we sort the users in descending order of $\mu_k(\theta = \rho)$, which results in an ordered list denoted by L_{user}. We set a variable m to count the number of groups, each of which uses the K&H scheduling with a fraction β_m. Denote $\mathcal{S}(m)$ the set that contains the users in group m.

We partition the K users recursively, starting from group $m = 1$. Each time when we form a new group m, we first remove the head element of list L_{user} and put it into an empty set $\mathcal{S}(m)$; if L_{user} is not empty, we again remove the head ν of list L_{user} and put ν into $\mathcal{S}(m)$; now we have two users and can apply the K&H scheduling to the two users; if the resulting channel usage is greater than or equal to that due to applying the RR scheduling to the two users, we move the user ν from $\mathcal{S}(m)$ back to the head of L_{user}, *i.e.*, the

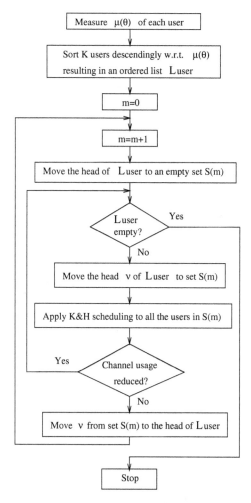

Figure 44: Flow chart of resource allocation for the heterogeneous case.

new user ν should not be added to the group m and group m will only have one user; otherwise, we recursively continue this procedure, that is, adding a new user and testing whether the resulting channel usage is reduced, i.e., $\beta_m(\mathcal{S}(m)) < \beta_m(\mathcal{S}'(m)) + r_s/\mu_\nu(\theta = \rho)$, where $\beta_m(\mathcal{S}(m))$ is the channel usage of set $\mathcal{S}(m)$ and set $\mathcal{S}'(m)$ is the difference $\mathcal{S}(m) - \{\nu\}$, and $r_s/\mu_\nu(\theta = \rho)$ is the channel usage for user ν if the RR scheduling is used. We continue the partitioning until L_{user} is empty.

In the process of partitioning, we determine ζ_m and β_m for scheduling as below. If group m has only one user, say user k, we only use the RR scheduling with $\zeta_m = r_s/\mu_k(\theta = \rho)$, and set $\beta_m = 0$; if group m has more than one user, we only use the K&H scheduling with the β_m obtained in the test for $\mathcal{S}(m)$, and set $\zeta_m = 0$. The joint K&H/RR scheduling is not used, in order to reduce complexity.

The outputs of the algorithm are 1) a partition of the K users, say, M groups, and 2) $\{\zeta_m, \beta_m\}$ $(m = 1, \cdots, M)$. After running the resource allocation algorithm, we do admission control by testing whether the total channel usage of the M groups is not greater than unity.

If the admission control is passed, we schedule each of the M groups as follows. If group m has only one user, we use the RR scheduling with ζ_m; if group m has more than one user, we apply the K&H scheduling with β_m to all the users in group m.

Our algorithm reduces complexity at the cost of optimality. Specifically, the resource allocation algorithm achieves $O(K \log K)$ complexity due to sorting. Note that we only have to try at most K tests in partitioning the K users. The performance of the algorithm may not be optimal but our simulations show that for practical range of Doppler rates, our algorithm improves the performance, compared to the RR scheduling.

Case 2: the K users have different $\{r_s, \rho\}$ and different channel statistics.

We classify the users into N classes so that the users in the same class have the same QoS pair $\{r_s, \rho\}$. Then apply the resource allocation algorithm to each class. The admission control is simply to check whether the total channel usage of the N classes is not greater than unity.

To summarize, given the fading channel and QoS of K users, we use the following procedure to achieve multiuser diversity gain with QoS provisioning:

1. Use the resource allocation algorithm to partition the K users and determine $\{\zeta_m, \beta_m\}$ that satisfies users' QoS.

2. If admission control is passed, provide the scheduler with ζ_m and β_m, for the RR or K&H scheduling.

5.4 Simulation Results

5.4.1 Simulation Setting

We simulate the system depicted in Figure 43, where K connections are set up and each mobile user is associated with one connection. Each connection is simulated as plotted in Figure 26. The data source of user k generates packets at a *constant* rate $r_s^{(k)}$. Generated packets are first sent to the (infinite) buffer at the transmitter. The head-of-line packet in the queue is transmitted over the fading channel at data rate $r_k(t)$. The fading channel has a random power gain $g_k(t)$. We use a fluid model, that is, the size of a packet is infinitesimal. In practical systems, the results presented here will have to be modified to account for finite packet sizes.

We assume that the transmitter has perfect knowledge of the current channel gains $g_k(t)$ in frame t. Therefore, it can use rate-adaptive transmissions, and ideal channel codes, to transmit packets without decoding errors. For the homogeneous case, under the joint scheduling, the transmission rate $r_k(t)$ of user k is equal to a fraction of its instantaneous capacity, as below,

$$r_k(t) = \begin{cases} (\zeta + \beta)c_k(t) & \text{if } k = \arg\max_{i \in \{1, \cdots, K\}} g_i(t); \\ \zeta c_k(t) & \text{otherwise.} \end{cases} \tag{120}$$

where the instantaneous channel capacity $c_k(t)$ is

$$c_k(t) = B_c \log_2(1 + g_k(t) \times P_0/\sigma_n^2) \tag{121}$$

where B_c denotes the channel bandwidth, and the transmission power P_0 and noise variance σ_n^2 are assumed to be constant. For the heterogeneous case, the rate $r_k(t)$ is computed within each class as described in Section 5.3.2.

The average SNR is fixed in each simulation run. We define r_{awgn} as the capacity of an equivalent AWGN channel, which has the same average SNR, *i.e.*,

$$r_{awgn} = B_c \log_2(1 + SNR_{avg}) \tag{122}$$

where $SNR_{avg} = E[g_k(t) \times P_0/\sigma^2] = P_0/\sigma^2$. We set $E[g_k(t)] = 1$.

The sample interval (frame length) T_s is set to 1 milli-second. Most simulation runs are 1000-second long; some simulation runs are 10000-second long in order to obtain good estimate of the actual delay-violation probability $Pr\{D(\infty) \geq D_{max}\}$ by the Monte Carlo method.

As described in Section 4.4.1, Rayleigh flat-fading voltage-gains $h_k(t)$ are generated by an AR(1) model as below. We first generate $\bar{h}_k(t)$ by

$$\bar{h}_k(t) = \kappa \times \bar{h}_k(t-1) + u_k(t), \tag{123}$$

133

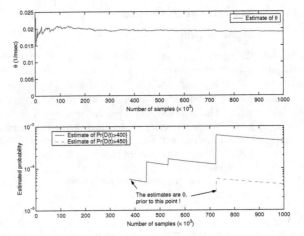

Figure 45: Convergence of estimates.

where $u_k(t)$ are i.i.d. complex Gaussian variables with zero mean and unity variance per dimension. Then, we normalize $\bar{h}_k(t)$ and obtain $h_k(t)$ by

$$h_k(t) = \bar{h}_k(t) \times \sqrt{\frac{1 - \kappa^2}{2}}. \tag{124}$$

For Ricean fading, the voltage-gains $h_k(t)$ are generated by adding a constant to Rayleigh-fading voltage-gains (see [104] for detail).

5.4.2 Performance Evaluation

We organize this section as follows. Section 5.4.2.1 shows the convergence of our estimation algorithm. In Section 5.4.2.2, we assess the accuracy of our QoS estimation (108). Section 5.4.2.3 investigates the effectiveness of the resource allocation scheme in QoS provisioning. In these sections (5.4.2.1 to 5.4.2.3), we only consider the homogeneous case, *i.e.*, all users have the same QoS requirements $\{r_s, \rho\}$ and also the same channel statistics. In Sections 5.4.2.4 through 5.4.2.6, we evaluate the performance of our scheduler in the homogeneous and the heterogeneous case, respectively.

134

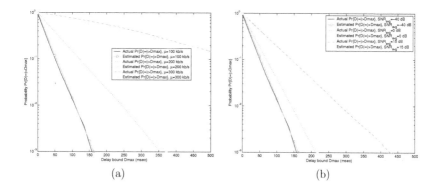

(a)   (b)

Figure 46: Actual and estimated delay-bound violation probability for (a) different source rates, and (b) different SNR_{avg}.

5.4.2.1 Convergence of Estimates

This experiment is to show the convergence behavior of estimates. We do simulations with the following parameters fixed: $r_{awgn} = 1000$ kb/s, $K = 10$, $\kappa = 0.8$, and $SNR_{avg} = -40$ dB.

Figure 45 shows the convergence of the estimate of θ ($\theta(\mu)$ for $\mu = 200$ kb/s) for the queue. It can be seen that the estimate of θ converges within 2×10^4 samples/frames (20 sec). The same figure shows the (lack of) convergence of direct (Monte Carlo) estimates of delay-bound-violation probabilities, measured for the same queue (the two probability estimates eventually converge to 10^{-3} and 10^{-4}, respectively). This precludes using the direct probability estimate to predict the user QoS, as alluded to in Section 5.1. The reason for the slow convergence of the direct probability estimate is that the K&H scheduling results in a user being allotted the channel in a bursty manner, and thus increases the correlation time of $D(t)$ substantially. Therefore, even 10^6 samples are not enough to obtain an accurate estimate of a probability as high as 10^{-3}.

5.4.2.2 Accuracy of Channel Estimation

The experiments in this section are to show that the estimated effective capacity can indeed be used to accurately predict QoS.

We do experiments under five different settings: 1) AR(1) Rayleigh fading channel with changing source rate and fixed SNR_{avg}, 2) AR(1) Rayleigh fading

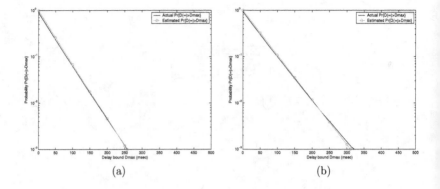

(a)   (b)

Figure 47: Actual and estimated delay-bound violation probability for (a) AR(2) channel, and (b) Ricean channel.

channel with changing SNR_{avg} and fixed source rate, 3) AR(2) Rayleigh fading channel, 4) Ricean fading channel, and 5) Nakagami-m fading channel (chi-distribution) [118, page 22].

Under the first setting, we do simulations with the following parameters fixed: $r_{awgn} = 1000$ kb/s, $K = 10$, $\kappa = 0.8$, and $SNR_{avg} = -40$ dB. By changing the source rate μ, we simulate three cases, $i.e.$, $\mu = 100$, 200, and 300 kb/s. Figure 46(a) shows the actual delay-bound violation probability $Pr\{D(\infty) > D_{max}\}$ vs. the delay bound D_{max}. From the figure, it can be observed that the actual delay-bound violation probability decreases exponentially with D_{max}, for all the cases. This confirms the exponential dependence shown in (108).

In addition, we use the estimation scheme, $i.e.$, Eqs. (36) through (40), to obtain an estimated θ; with the resulting θ, we predict the probability $Pr\{D(\infty) > D_{max}\}$ (using (108)). As shown in Figure 46(a), the estimated $Pr\{D(\infty) > D_{max}\}$ is quite close to the actual $Pr\{D(\infty) > D_{max}\}$. This demonstrates that our estimation is accurate, which justifies the use of (111) by the resource allocation algorithm to guarantee QoS.

Notice that the (negative) slope of the $Pr\{D(\infty) > D_{max}\}$ plot increases with the decrease of the source rate μ. This is because the smaller the source rate, the smaller the probability of delay-bound violation, resulting in a sharper slope ($i.e.$, a larger decaying rate θ).

Under the second setting, we do simulations with the following parameters fixed: $r_{awgn} = 1000$ kb/s, $K = 10$, $\kappa = 0.8$, and $\mu = 100$ kb/s. By

136

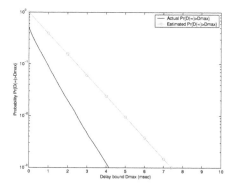

Figure 48: Actual and estimated delay-bound violation probability for Nakagami-m channel ($m = 32$).

changing SNR_{avg}, we simulate three cases, *i.e.*, SNR_{avg} = -40, 0, and 15 dB. Figure 46(b) shows that the conclusions drawn from the first set of experiments still hold. Thus, our estimation scheme gives consistent performance over different SNRs also.

In the third setting, AR(2) Rayleigh fading voltage-gains $h_k(t)$ are generated as below:

$$h_k(t) = \kappa_1 \times h_k(t-1) + \kappa_2 \times h_k(t-2) + v_k(t), \qquad (125)$$

where $v_k(t)$ are zero-mean i.i.d. complex Gaussian variables. The parameters of the simulation are $r_{awgn} = 1000$ kb/s, $\kappa_1 = 0.7$, $\kappa_2 = 0.2$, $K = 10$, $\mu = 100$ kb/s and $SNR_{avg} = -40$ dB. Figure 47(a) shows that the conclusions drawn from the first set of experiments still hold. Thus, our estimation scheme consistently predicts the QoS metric under different autoregressive channel fading models.

Under the fourth setting, the parameters of the simulation are Ricean factor[2] = 7 dB, $r_{awgn} = 1000$ kb/s, $K = 10$, $\kappa = 0.8$, and $\mu = 100$ kb/s. Figure 47(b) shows that the conclusions drawn from the first set of experiments also hold for Ricean fading channels.

In the fifth setting, Nakagami-m fading power gains $g_k(t)$ are generated as

---

[2]Ricean factor is defined as the ratio between the deterministic signal power A^2 and the variance of the multipath $2\sigma_m^2$, *i.e.*, Ricean factor = $A^2/(2\sigma_m^2)$.

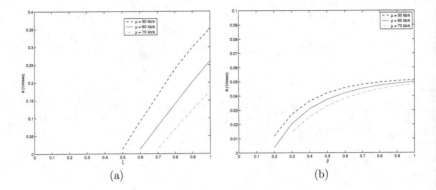

Figure 49: (a) θ vs. ζ and (b) θ vs. β.

below:

$$g_k(t) = \sum_{i=1}^{m} \hat{g}_i(t), \tag{126}$$

where m is the parameter of Nakagami-m distribution and takes values of positive integers, $\hat{g}_i(t)$ are AR(1) Rayleigh fading power gains. The parameters of the simulation are $m = 32$, $r_{awgn} = 1000$ kb/s, $K = 10$, $\kappa = 0.8$, $SNR_{avg} = -40$ dB, and $\mu = 90$ kb/s. Figure 48 shows that the estimate does not give a good agreement with the actual $Pr\{D(\infty) > D_{max}\}$. The reason is that the high diversity in high-order Nakagami fading models averages out the randomness in the fading process. The higher diversity a fading channel has, the more like an AWGN channel the fading channel is. It is known that for an AWGN channel, the actual $Pr\{D(\infty) > D_{max}\}$ does not decay exponentially with D_{max} but takes values of 0 or 1. Therefore, the higher diversity a fading channel possesses, the less accurate the exponential approximation (108) is, hence the less accurate the estimate is.

In summary, the results for Rayleigh/Ricean flat-fading channels have shown the exponential behavior of the actual $Pr\{D(\infty) > D_{max}\}$ and the accurateness of our estimation. We caution however that such a strong agreement between the estimate and the actual QoS may not occur in all situations with practical values of D_{max} (although the theory predicts the agreement asymptotically for large D_{max}). We have shown that in the case of high-diversity channel fading models (*e.g.*, high-order Nakagami fading models), the estimation is not accurate.

138

5.4.2.3 Effectiveness of Resource Allocation in QoS Provisioning

The experiments here are to show that a QoS pair $\{r_s, \rho\}$ can be achieved (within limits) by choosing ζ or β appropriately. In the experiments, we fix the following parameters: $K = 10$, $\kappa = 0.8$, and $SNR_{avg} = -40$ dB. We simulate three data rates, *i.e.*, $\mu = 50$, 60, and 70 kb/s, respectively. We do two sets of experiments: one for the RR scheduling and the other for the K&H scheduling.

In the first set of experiments, only the RR scheduling is used; we change ζ from 0.1 to 1 and estimate the resulting θ for a given μ, using Eqs. (36) through (40). Figure 49(a) shows that θ increases with ζ. Thus, Figure 49 can be used to allot ζ to a user to satisfy its QoS requirements when using RR scheduling.

In the second set of experiments, only the K&H scheduling is used; we change β from 0.1 to 1 and estimate the resulting θ, for a given μ. Figure 49(b) shows that θ increases with the increase of β, and thus the figure can be used to allot β to a user to satisfy its QoS requirements when using K&H scheduling.

5.4.2.4 Performance Gains of Scheduling: Homogeneous Case

Under the setting of identical QoS requirements $\{r_s, \rho\}$ and i.i.d. channel gain processes, the experiments here demonstrate the performance gain of joint scheduling over RR scheduling, using the optimum $\{\zeta, \beta\}$ values specified by the resource allocation algorithm, *i.e.*, Eqs. (110)–(113). In particular, the experiments show that for loose delay constraints, the large capacity gains promised by the K&H scheme can indeed be approached.

To evaluate the performance of the scheduling schemes under different SNRs and different Doppler rates (*i.e.*, different κ), we simulate three cases: 1) $\kappa = 0.8$, and $SNR_{avg} = -40$ dB, 2) $\kappa = 0.8$, and $SNR_{avg} = 15$ dB, and 3) $\kappa = 0.95$, and $SNR_{avg} = -40$ dB. In all the experiments, we set $r_{awgn} = 1000$ kb/s and $K = 10$.

In Figure 50, we plot the function $\theta(\mu)$ achieved by the joint, K&H, and RR schedulers, for a range of source rate μ, when the entire frame is used (*i.e.*, $K\zeta + \beta = 1$). The function $\theta(\mu)$ in the figure is obtained by the estimation scheme, *i.e.*, Eqs. (36) through (40). In the case of the joint scheduling, each point in the figure corresponds to a specific optimum $\{\zeta, \beta\}$, while for the RR and the K&H scheduling, we set $K\zeta = 1$ and $\beta = 1$ respectively. The curve of $\theta(\mu)$ can be directly used to check for feasibility of a QoS pair $\{r_s, \rho\}$, by checking whether $\theta(r_s) > \rho$ is satisfied. Furthermore, for a given θ, the ratio of $\mu(\theta)$ of the joint scheduler to the $\mu(\theta)$ of the RR scheduler (both obtained from the figure), represents the delay-constrained **capacity gain** that can be achieved by using joint scheduling.

139

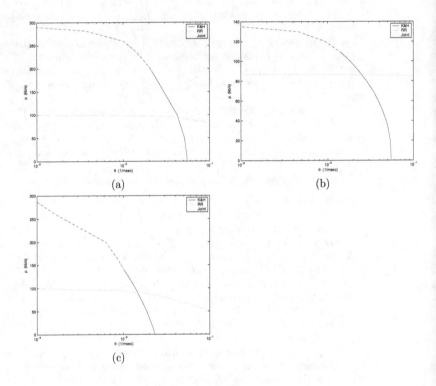

Figure 50: $\theta(\mu)$ vs. μ for (a) $\kappa = 0.8$, $SNR_{avg} = -40$ dB, (b) $\kappa = 0.8$, $SNR_{avg} = 15$ dB, and (c) $\kappa = 0.95$, $SNR_{avg} = -40$ dB.

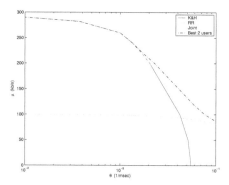

Figure 51: $\theta(\mu)$ vs. μ for splitting the channel between the best two users.

Four important observations can be made from the figure. First, the range of θ can be divided into three segments: small, medium, and large θ, which correspond to three categories of the QoS constraints: loose-delay, medium-delay, and tight-delay requirements. For small θ, our joint scheduler achieves substantial gain, $e.g.$, approximately $\sum_{k=1}^{K} \frac{1}{k}$ capacity gain for Rayleigh fading channels at low SNR. For example, in Figure 50(a), when $\theta = 0.001$, the capacity gain for the joint scheduler is 2.9, which is close to $\sum_{k=1}^{10} \frac{1}{k} = 2.929$. For medium θ, our joint scheduler also achieves gain. For example, in Figure 50(a), when $\theta = 0.01$, the capacity gain for the joint scheduler is 2.6. For large θ, such as $\theta = 0.1$, our joint scheduler does not give any gain. Thus, the curve of $\theta(\mu)$ shows the range of θ (delay constraints) for which a K&H type scheme can provide a performance gain. When the scheduler is provided with the optimum $\{\zeta, \beta\}$ values, the QoS pair $\{\mu, \theta(\mu)\}$ guaranteed to a user is indeed satisfied; the simulation result that shows this fact, is similar to Figure 46, and therefore, is not shown.

Second, we observe that the joint scheduler has a larger effective capacity than both the K&H and the RR for a rather small range of θ. Therefore, in practice, it may be sufficient to use either K&H or RR scheduling, depending on whether θ is small or large respectively, and dispense with the more complicated joint scheduling. However, we have designed more sophisticated joint schedulers, such as splitting the channel between the best two users in every slot, which perform substantially better than either K&H and RR scheduling, for medium values of θ. This is shown in Figure 51. This indicates that more sophisticated joint schedulers may squeeze out more channel capacity gain. We leave this for future study.

141

Third, the figure of $\theta(\mu)$ can be used to satisfy the QoS constraint (111), even though it only represent the $K\zeta + \beta = 1$ case, as follows. For the QoS pair $\{r_s, \rho\}$, we compute the ratio $\lambda \doteq \frac{r_s}{\mu(\theta=\rho)}$ using the $\mu(\theta)$ function in the figure. Suppose the $\mu(\theta = \rho)$ point in the figure corresponds to the optimum pair $\{\bar{\zeta}, \bar{\beta}\}$. Since we have the relation $\theta_{\bar{\zeta},\bar{\beta}}(\mu) = \theta_{\lambda\bar{\zeta},\lambda\bar{\beta}}(\lambda\mu)$, i.e., Eq. (114), we assert that instead of using the entire frame (as in the figure), if we use a total fraction λ of the frame, then we can achieve the desired QoS $\{r_s, \rho\}$. The joint scheduler then needs to use the $\{\lambda\bar{\zeta}, \lambda\bar{\beta}\}$ pair to do RR and K&H scheduling respectively. This indicates a compelling advantage of our QoS provisioning scheme over direct-measurement based schemes, which require experiments for different λ, even if the ratio ζ/β is fixed.

Fourth, we observe that if θ is larger than a certain value, the corresponding data rate $\mu(\theta)$ achieved by the K&H, approaches zero. This is because the probability that a user will be allowed to transmit is $1/K$ since there are K identical users share the channel; hence, on average, a user has to wait $K - 1$ frames before its next packet is scheduled for transmission, making a guarantee of tight-delay requirements (large θ) impossible.

5.4.2.5 Performance Improvement due to Partitioning

This experiment demonstrates the performance improvement due to partitioning of users as mentioned in Section 5.3.1.3. In particular, the experiment indicates the trade-off between performance improvement and complexity.

Again, the setting of this experiment is i.i.d. channel gain processes and identical QoS requirements $\{r_s, \rho\}$. We divide K users into groups and see how much gain can be achieved, compared with non-partitioning. The parameters of the experiment is the following: $r_{awgn} = 1000$ kb/s, $\kappa = 0.8$, $SNR_{avg} = -40$ dB, $r_s = 100$ kb/s, $\rho = 0.03$.

The simulation result shows that the channel can support at most 13 users if the set of users is not partitioned. If users are allowed to be partitioned, solving Eqs. (115)–(119), we find the maximum number of users that the channel can support increases to 16, where the 16 users are partitioned into two groups, each of which consists of 8 users. Note that this performance improvement is at the cost of complexity of solving Eqs. (115)–(119).

5.4.2.6 Performance Gains of Scheduling: Heterogeneous Case

The experiments here show the performance gain of the K&H/RR scheduling over the RR scheduling, using the $\{\zeta_m, \beta_m\}$ values specified by the resource allocation algorithm in Figure 44, under the setting of identical QoS requirements $\{r_s, \rho\}$ and non-identical, independent channel gain processes.

142

We do two sets of experiments. The first set of experiments is to show the performance gain of the algorithm in Figure 44 under different number of users, while the second set of experiments is to show the performance gain of the algorithm under different types of channels. In the experiments, we fix the following parameters: $r_{awgn} = 1000$ kb/s, $SNR_{avg} = -40$ dB, $r_s = 30$ kb/s, $\rho = 0.01$.

The first set of experiments is done under three different settings: 1) $K = 10$, 2) $K = 19$, 3) $K = 37$. We use AR(1) Rayleigh fading channel and each user has different κ, i.e., different Doppler rate. We let κ change from 0.6 to 0.99 with equal spacing for the three settings, that is, user 1 has $\kappa = 0.6$, the last user (10th, 19th, or 37th user) has $\kappa = 0.99$, and other users' κ are determined by equal spacing between 0.6 and 0.99.

In the first setting ($K = 10$), the resource allocation algorithm in Figure 44 results in a partition with two groups, where the number of users in each group are $K_1 = 8$ and $K_2 = 2$, respectively. The total channel usage under the K&H/RR scheduling is 0.195, while the total channel usage under the RR scheduling is 0.319.

In the second setting ($K = 19$), the resource allocation algorithm leads to a partition with three groups, where $K_1 = 14$, $K_2 = 4$, and $K_3 = 1$. The total channel usage under the K&H/RR scheduling is 0.344, while the total channel usage under the RR scheduling is 0.597.

In the third setting ($K = 37$), the resource allocation algorithm obtains a partition with five groups, where $K_1 = 27$, $K_2 = 6$, $K_3 = 2$, $K_4 = 1$, and $K_5 = 1$. The total channel usage under the K&H/RR scheduling is 0.69, while the total channel usage under the RR scheduling is 1.15, which is rejected by admission control.

The objective of these experiments is to see the performance gain of the algorithm in Figure 44 under different number of users.

The second set of experiments is done under the following setting. We have $K = 10$ users in the system. Among the ten users, four users have AR(1) Rayleigh fading channels and each of them has different κ. We let κ change from 0.6 to 0.99 with equal spacing, that is, user 1 has $\kappa = 0.6$, the fourth user has $\kappa = 0.99$, and other users' κ are determined by equal spacing between 0.6 and 0.99. The other six users have AR(2) Rayleigh fading channels, specified by (125). Table 5 lists the parameters for the AR(2) Rayleigh fading channels of the six users. From the simulation, the resource allocation algorithm obtains a partition with two groups, where $K_1 = 9$ and $K_2 = 1$. The total channel usage under the K&H/RR scheduling is 0.1842, while the total channel usage under the RR scheduling is 0.3215.

Hence, our resource allocation algorithm in Figure 44 achieves smaller channel usage than that using the RR scheduling; as a result, the system can admit

143

Table 5: Parameters for AR(2) Rayleigh fading channels.

	User 1	User 2	User 3	User 4	User 5	User 6
κ_1	0.8	0.75	0.7	0.6	0.5	0.45
κ_2	0.1	0.05	0.2	0.2	0.3	0.4

more users.

In summary, the K&H/RR scheduler achieves performance gain when delay requirements are not very tight, while yet guaranteeing QoS at any delay requirement.

5.5   Summary

In this chapter, we examined the problem of QoS provisioning for K users sharing a single time-slotted fading downlink channel. We developed simple and efficient schemes for admission control, resource allocation, and scheduling, to obtain a gain in delay-constrained capacity. Multiuser diversity obtained by the well-known K&H scheduling is the key that gives rise to this performance gain. However, the unique feature of this chapter is explicit support of the statistical QoS requirement $\{r_s, D_{max}, \varepsilon\}$, for channels utilizing K&H scheduling. The concept of effective capacity is the key that explicitly guarantees the QoS. Thus, this chapter combines crucial ideas from the areas of communication theory and queueing theory to provide the tools to increase capacity and yet satisfy QoS constraints. The statistical QoS requirement is satisfied by the channel assignments $\{\zeta, \beta\}$, which are determined by the resource allocation module at the admission phase. Then, the joint scheduler uses the channel assignments $\{\zeta, \beta\}$ in scheduling data at the transmission phase, with guaranteed QoS. Simulation results have shown that our approach can substantially increase the delay-constrained capacity of a fading channel, compared with the RR scheduling, when delay requirements are not very tight.

In the next chapter, we will extend the work in this chapter and focus on the design of scheduling, admission control and resource allocation, for *multiple users* sharing *multiple channels*.

144

CHAPTER 6

QOS PROVISIONING: MULTIPLE SHARED CHANNELS

6.1 Introduction

In Chapter 5, we recognized that the key difficulty in explicit QoS provisioning, is the lack of a method that can easily relate the control parameters of a QoS provisioning system to the QoS measures, and hence we proposed an approach to simplify the task of explicit provisioning of QoS guarantees. Specifically, we simplify the design of joint K&H/RR scheduler by shifting the burden to the resource allocation mechanism. Furthermore, we are able to solve the resource allocation problem efficiently, thanks to the method of effective capacity we developed in Chapter 4. Effective capacity captures the effect of channel fading on the queueing behavior of the link, using a computationally simple yet accurate model, and thus, is the critical device we need to design an efficient resource allocation mechanism.

Chapter 5 presented QoS provisioning mechanisms for multiple users sharing *one* channel. This chapter is intended to extend our QoS provisioning mechanisms to the setting of multiple users sharing *multiple* parallel channels, by utilizing both multiuser diversity and frequency diversity. Due to the frequency diversity inherent in multiple wireless channels, the joint K&H/RR scheduler in the new setting can achieve higher capacity gain than that in Chapter 5. Moreover, when users' delay requirements are stringent, wherein the joint K&H/RR reduces to the RR scheduling, the high capacity gain associated with K&H scheduling vanishes. To squeeze out more capacity in this case, a possible solution is to design a scheduler, which dynamically selects the best channel among multiple channels for a user to transmit. In other words, this scheduler is intended to find a channel-assignment schedule, at each time-slot, which minimizes the channel usage while yet satisfying users' QoS requirements.

We formulate this scheduling problem as a linear program, in order to avoid the 'curse of dimensionality' associated with optimal dynamic programming solutions. The key idea that allows us to do this, is what we call the 'Reference Channel' approach, wherein the QoS requirements of the users, are captured by

145

resource allocation (channel assignments). The scheduler obtained, as a result of the Reference Channel approach, is sub-optimal. Therefore, we analyze the performance of this scheduler, by comparing its performance gain with a bound we derived. We show by simulations, that the performance of our sub-optimal scheduler is quite close to the bound. This demonstrates the effectiveness of our scheduler. The performance gain is obtained, as a result of dynamically choosing the best channel to transmit.

The remainder of this chapter is organized as follows. In Section 6.2, we present efficient QoS provisioning mechanisms and show how to use multiuser diversity and frequency diversity to achieve a capacity gain while yet satisfying QoS constraints. Section 6.3 describes our reference-channel-based scheduler that provides a performance gain when delay requirements are tight. In Section 6.4, we present the simulation results that illustrate the performance improvement of our scheme over that in Chapter 5. Section 6.6 summarizes the chapter.

6.2 QoS Provisioning with Multiuser Diversity and Frequency Diversity

This section is organized as below. Section 6.2.1 describes the assumptions and the QoS provisioning architecture we use. In Section 6.2.2, we present efficient schemes for guaranteeing QoS.

6.2.1 Architecture

Figure 52 shows the architecture for transporting multiuser traffic over time-slotted fading channels. A cellular wireless network is assumed, and the downlink is considered, where a base station transmits data over N parallel, independent channels to K mobile user terminals, each of which requires certain QoS guarantees. The channel fading processes of the users are assumed to be stationary, ergodic and independent of each other. (Compare Figure 52 to Figure 42 in Chapter 5, which assumes a single channel only.) A single cell is considered, and interference from other cells is modelled as background noise with constant variance. We assume a block fading channel model [13], which assumes that user channel gains are constant over a time duration of length T_s (T_s is assumed to be small enough that the channel gains are constant, yet large enough that ideal channel codes can achieve capacity over that duration). Therefore, we partition time into 'frames' (indexed as $t = 0, 1, 2, \ldots$), each of length T_s. Thus, each user k has time-varying channel power gains $g_{k,n}(t)$, for each of the N independent channels, which vary with the frame index t. Here $n \in \{1, 2, \ldots, N\}$ refers to the nth channel. The base station is assumed to

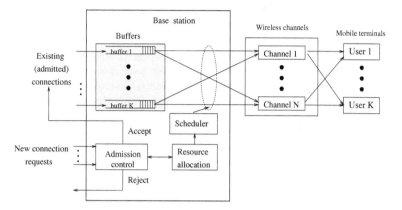

Figure 52: QoS provisioning architecture for multiple channels.

know the current and past values of $g_{k,n}(t)$. The capacity of the nth channel for the kth user, $c_{k,n}(t)$, is

$$c_{k,n}(t) = \log_2(1 + g_{k,n}(t) \times P_0/\sigma^2) \qquad \text{bits/symbol} \qquad (127)$$

where the transmission power P_0 and noise variance σ^2 are assumed to be constant and equal for all users. We divide each frame of length T_s into infinitesimal time slots, and assume that the same channel n can be shared by several users, in the same frame. This is illustrated in Figure 52, where data from buffers 1 to K can be simultaneously transmitted over channel 1. Further, we assume a *fluid model* for packet transmission, where the base station can allot *variable fractions* of a channel frame to a user, over time. Thus, the model differs from that in Chapter 5, in that *multiple parallel* time-varying channels are assumed here. The system described above could be, for example, an idealized FDMA-TDMA[1] system, where the N parallel, independent channels represent N frequencies, which are spaced apart (FDMA), and where the frame of each channel consists of TDMA time slots which are infinitesimal. Note that in a practical FDMA-TDMA system, there would be a finite number of finite-length time slots in each frame, rather than the infinite number of infinitesimal time slots, assumed here.

As shown in Figure 52, our QoS provisioning architecture consists of three components, namely, admission control, resource allocation, and scheduling.

---

[1]FDMA is frequency-division multiple access and TDMA is time-division multiple access.

When a new connection request comes, we first use a resource allocation algorithm to compute how much resource is needed to support the requested QoS. Then the admission control module checks whether the required resource can be satisfied. If yes, the connection request is accepted; otherwise, the connection request is rejected. For admitted connections, packets destined to different mobile users[2] are put into separate queues. The scheduler decides, in each frame t, how to schedule packets for transmission, based only on the *current* channel gains $g_{k,n}(t)$ and the amount of resource allocated.

In the next section, we describe QoS provisioning schemes under the aforementioned architecture.

6.2.2 QoS Provisioning Schemes

6.2.2.1 Scheduling

As explained in Section 6.1, we simplify the scheduler, by shifting the burden of guaranteeing users' QoS to resource allocation. Therefore, our scheduler is a simple combination of K&H and RR scheduling.

We first explain K&H and RR scheduling separately. In any frame t, the K&H scheduler transmits the data of the user with the largest gain $g_{k,n}(t)$ ($k = 1, 2, \cdots, K$), for *each* channel n. However, the QoS of a user may be satisfied by using only a fraction of the frame $\beta \leq 1$. Therefore, it is the function of the resource allocation algorithm to allot the minimum required β to the user. This will be described in Section 6.2.2.2. It is clear that K&H scheduling attempts to utilize multiuser diversity to maximize the throughput of each channel. Compared to the K&H scheduling over single channel as described in [158], the K&H scheduling here achieves higher throughput when delay requirements are loose. This is because, for fixed ratio[3] N/K, as the number of channel N increases, the number of users K increases, resulting in a larger capacity gain, which is approximately $\sum_{k=1}^{K} 1/k$.

On the other hand, for *each* channel n, the RR scheduler allots to every user k, a fraction
$\zeta \leq 1/K$ of *each* frame, where ζ again needs to be determined by the resource allocation algorithm. Thus RR scheduling attempts to provide tight QoS guarantees, at the expense of decreased throughput, in contrast to K&H scheduling. Compared to the RR scheduling over single channel as described in [158], the RR scheduling here utilizes frequency diversity (each user's data simultaneously transmitted over multiple channels), thereby increasing effective

---

capacity when delay requirements are tight.

Our scheduler is a joint K&H/RR scheme, which attempts to maximize the throughput, while yet providing QoS guarantees. In each frame t and for each channel n, its operation is the following. First, find the user $k^*(n,t)$ such that it has the largest channel gain among all users, for channel n. Then, schedule user $k^*(n,t)$ with $\beta + \zeta$ fraction of frame t in channel n; schedule each of the other users $k \neq k^*(n,t)$ with ζ fraction of frame t in channel n. Thus, for each channel, a fraction β of the frame is used by K&H scheduling, while simultaneously, a total fraction $K\zeta$ of the frame is used by RR scheduling. Then, for each channel n, the total usage of the frame is $\beta + K\zeta \leq 1$.

6.2.2.2   Admission Control and Resource Allocation

The scheduler described in Section 6.2.2.1 is simple, but it needs the frame fractions $\{\beta, \zeta\}$ to be computed and reserved. This function is performed at the admission control and resource allocation phase.

Since we only consider the homogeneous case, without loss of generality, denote $\alpha_{\zeta,\beta}(u)$ the effective capacity function of user $k = 1$ under the joint K&H/RR scheduling (henceforth called 'joint scheduling'), with frame shares ζ and β respectively, i.e., denote the capacity process allotted to user 1 by the joint scheduler as the process $r(t)$ and then compute $\alpha_{\zeta,\beta}(u)$ using (28). The corresponding QoS exponent function $\theta_{\zeta,\beta}(\mu)$ can be found via (30). Note that since the capacity process $r(t)$ depends on the number of users K and the number of channels N, $\theta_{\zeta,\beta}(\mu)$ is actually a function of K and N. However, since we assume K and N are fixed, there is no need to put the extra arguments K and N in the function $\theta_{\zeta,\beta}(\mu)$. With this simplified notation, the admission control and resource allocation scheme for users requiring the QoS pair $\{r_s, \rho\}$ is given as below,

$$\underset{\{\zeta,\beta\}}{\text{minimize}} \quad K\zeta + \beta \tag{128}$$

$$\text{subject to} \quad \theta_{\zeta,\beta}(r_s) \geq \rho, \tag{129}$$

$$K\zeta + \beta \leq 1, \tag{130}$$

$$\zeta \geq 0, \qquad \beta \geq 0. \tag{131}$$

The minimization in (128) is to minimize the total frame fraction used. (129) ensures that the QoS pair $\{r_s, \rho\}$ of each user is feasible. Furthermore, Eqs. (129)–(131) also serve as an admission control test, to check availability of resources to serve this set of users. Since we have the relation $\theta_{\zeta,\beta}(\mu) = \theta_{\lambda\zeta,\lambda\beta}(\lambda\mu)$ (Proposition 5.2), we only need to measure the $\theta_{\zeta,\beta}(\cdot)$ functions for different ratios of ζ/β.

To summarize, given N fading channels and QoS of K homogeneous users, we use the following procedure to achieve multiuser/frequency diversity gain with QoS provisioning:

1. Estimate $\theta_{\zeta,\beta}(\mu)$, directly from the queueing behavior, for various values of $\{\zeta, \beta\}$.

2. Determine the optimal $\{\zeta, \beta\}$ pair that satisfies users' QoS, while minimizing frame usage.

3. Provide the joint scheduler with the optimal ζ and β, for simultaneous RR and K&H scheduling, respectively.

It can be seen that the above joint K&H/RR scheduling, admission control and resource allocation schemes utilize both multiuser diversity and frequency diversity. We will show, in Section 6.4, that such a QoS provisioning achieves higher effective capacity than the one described in Chapter 5, which utilizes multiuser diversity only.

On the other hand, we observe that when users' delay requirements are stringent, the joint K&H/RR reduces to the RR scheduling (fixed slot assignment) (see Figure 55). Then the high capacity gain associated with the K&H scheduling cannot be achieved (see Figure 55). Can the scheduling be modified, so that even with stringent delay requirements, gains over simple RR scheduling can be achieved? To answer this question, we provide an analogy to diversity techniques used in physical layer designs. The careful reader may notice that the RR scheduler proposed in Section 6.2.2.1 has a similar flavor to equal power distribution used in multichannel transmission, since the RR scheduler equally distributes the traffic of a user over multiple channels in each frame. Since transmitting over the best channel often achieves better performance than equal power distribution, one could ask whether choosing the best channel for each user to transmit (as opposed to choosing the best user for each channel as in Section 6.2.2.1), would bring about performance gain in the case of tight delay requirements. This is the motivation of designing a reference-channel-based scheduler for tight delay requirements, which we present next.

6.3 Reference-channel-based Scheduling

Section 6.2 basically extends the K&H/RR scheduling technique of Chapter 5, to the case with multiple parallel channels. The drawback of this straightforward extension was that, although the capacity gain is high for loose delay requirements (see Section 6.4.2.2), the gain vanishes when delay requirements become stringent. This section therefore proposes a scheduler, which squeezes

150

more out of frequency diversity, to provide capacity gains under stringent delay requirements.

This section is organized as follows. We first formulate the downlink scheduling problem in Section 6.3.1. Then in Section 6.3.2, we propose a reference channel approach to addressing the problem and with this approach we design the scheduler by posing it as a linear program. In Section 6.3.3, we investigate the performance of the scheduler.

6.3.1 The Problem of Optimal Scheduling

The scheduling problem is to find, for each frame t, the set of $\{w_{k,n}(t)\}$ that minimizes the time-averaged expected channel usage $\frac{1}{\tau} \sum_{t=0}^{\tau-1} \mathbf{E}[\sum_{k=1}^{K} \sum_{n=1}^{N} w_{k,n}(t)]$ (where τ is the connection life time), given the QoS constraints, as below,

$$\underset{\{w_{k,n}(t)\}}{\text{minimize}} \quad \frac{1}{\tau} \sum_{t=0}^{\tau-1} \mathbf{E}[\sum_{k=1}^{K} \sum_{n=1}^{N} w_{k,n}(t)] \tag{132}$$

$$\text{subject to} \quad Pr\{D_k(\infty) \geq D_{max}^{(k)}\} \leq \varepsilon_k, \quad \text{for a fixed rate } r_s^{(k)}, \quad \forall k \tag{133}$$

$$\sum_{k=1}^{K} w_{k,n}(t) \leq 1, \quad \forall n, \forall t \tag{134}$$

$$w_{k,n}(t) \geq 0, \quad \forall k, \forall n, \forall t \tag{135}$$

The constraint (133) represents statistical QoS constraints, that is, each user k specifies its QoS by a triplet $\{r_s^{(k)}, D_{max}^{(k)}, \varepsilon_k\}$, which means that each user k, transmitting at a fixed data rate $r_s^{(k)}$, requires that the probability of its steady-state packet delay $D_k(\infty)$ exceeding the delay bound $D_{max}^{(k)}$, is not greater than ε_k. The constraint (134) arises because the total usage of any channel n cannot exceed unity. The intuition of the formulation (132) through (135) is that, the less is the channel usage in supporting QoS for the K users, the more is the bandwidth available for use by other data, such as Best-Effort or Guaranteed Rate traffic [49].

We call any scheduler, which achieves the minimum in (132), as the *optimal scheduler*. To meet the statistical QoS requirements of the K users, an optimal scheduler needs to keep track of the queue length, for each user, using a state variable. It would make scheduling decisions (*i.e.*, allocation of $\{w_{k,n}(t)\}$), based on the current state. Dynamic programming often turns out to be a natural way to solve such an optimization problem [30, 43]. However, the dimensionality of the state variable is typically proportional to the number of users (at least), which results in very high (exponential in number of users)

151

complexity for the associated dynamic programming solution [8]. Simpler approaches, such as [4], which use the state variable sub-optimally, do not enforce a given QoS, but rather seek to optimize some form of a QoS parameter.

This motivates us to seek a simple (sub-optimal) approach, which can enforce the specified QoS constraints explicitly, and yet achieve an efficient channel usage. This idea is elaborated in the next section.

6.3.2 'Reference Channel' Approach to Scheduling

The key idea in the scheduler design is to specify the QoS constraints, using (what we call) the 'Reference Channel' approach. In the original optimal scheduling problem (132), the statistical QoS constraints (133) are specified by triplets $\{r_s^{(k)}, D_{max}^{(k)}, \varepsilon_k\}$. However, we map these constraints into a new form, based on the actual time-varying channel capacities of the K users. To elaborate, we assume that the base station can measure the statistics of the time-varying channel capacities (specifically, the QoS exponent function $\theta(\mu)$). Further, it is assumed that an appropriate admission control and resource allocation algorithm (such as that in Section 6.2.2.2), allots a fraction $\xi_{k,n}$ ($\xi_{k,n}$ are real numbers in the interval $[0, 1]$) of channel n, to user k, for the duration of the connection time. In other words, the key idea of the admission control and resource allocation algorithm is that, if a given user k were allotted the *fixed* channel assignment $\{\xi_{k,n}\}$ during the entire connection period, then the time-varying capacity $\sum_{n=1}^{N} \xi_{k,n} c_{k,n}(t)$, which it would obtain, would be sufficient to fulfill its QoS requirements specified by $\{r_s^{(k)}, D_{max}^{(k)}, \varepsilon_k\}$. A necessary condition on $\xi_{k,n}$ is that,

$$\sum_{k=1}^{K} \xi_{k,n} \leq 1, \qquad \forall \, n \tag{136}$$

Thus, our approach shifts the complexity of satisfying the QoS requirements (133), from the scheduler to the admission control algorithm, which needs to ensure that its choice of channel assignment $\{\xi_{k,n}\}$, meets the QoS requirements of all the users. Since the QoS constraint (133) is embedded in the channel assignment $\{\xi_{k,n}\}$, hence we call our approach to scheduling as a 'Reference Channel' approach. The careful reader may note a similarity of this approach, to other virtual reference approaches [162, 169], which are used to handle *source randomness* in wireline scheduling. Our motivation, on the other hand, is to handle *channel randomness* in wireless scheduling. This point is discussed in more detail in Section 6.5.

Thus, with the QoS constraints embedded in the $\{\xi_{k,n}\}$, the QoS constraint

152

(133) can be replaced by the specific set of constraints,

$$\sum_{n=1}^{N} w_{k,n}(t) c_{k,n}(t) \geq \sum_{n=1}^{N} \xi_{k,n} c_{k,n}(t), \quad \forall \, k \qquad (137)$$

Note that the channel fractions $w_{k,n}(t)$ and $\xi_{k,n}$ perform different functions. The fractions $w_{k,n}(t)$ are assigned by a *scheduler*, depending on the channel gains it observes, and they specify the actual fractions of the N channel frames used by different users at time t. Thus, they will (in general) vary with time. On the other hand, the fractions $\xi_{k,n}$ are assigned by an *admission control and resource allocation algorithm*, and they represent the channel resources *reserved* for different users, rather than the actual fractions of the N channel frames *used* by the users. Thus, $\xi_{k,n}$ are fixed during the life time of a connection. Note that setting $w_{k,n}(t) = \xi_{k,n}, \forall \, t$ ensures feasibility of (137) under all circumstances. This is simply the RR scheduling of Section 6.2.2.1! However, the enhanced scheduler we propose can satisfy (137), while (hopefully) also provide a capacity gain by minimizing the channel usage $\sum_{k=1}^{K} \sum_{n=1}^{N} w_{k,n}(t)$.

It is clear that (137) ensures that in every frame t, the scheduler will allot each user k a capacity, which is not less than the capacity specified by the $\xi_{k,n}$. Thus, a scheduler that satisfies (137) is guaranteed to satisfy the QoS requirements of all the K users. However, in the process of replacing the QoS constraint (133), by the constraint (137), we have conceivably tightened the constraints on the scheduler (since the latter constraint needs to be at least as tight as the former), which means that the scheduler we will derive will be sub-optimal, with respect to the optimal scheduler (132) through (135). However, as will be shown, this modification results in a simpler scheduler, which achieves a performance close to a bound we derived.

To summarize, we derive a sub-optimal scheduler, which we call Reference Channel (RC) scheduler, based on the optimization problem below: for each frame t,

$$\underset{\{w_{k,n}(t)\}}{\text{minimize}} \quad \sum_{k=1}^{K} \sum_{n=1}^{N} w_{k,n}(t) \qquad (138)$$

$$\text{subject to} \quad \sum_{n=1}^{N} w_{k,n}(t) c_{k,n}(t) \geq \sum_{n=1}^{N} \xi_{k,n} c_{k,n}(t), \quad \forall \, k \qquad (139)$$

$$\sum_{k=1}^{K} w_{k,n}(t) \leq 1, \quad \forall \, n \qquad (140)$$

$$w_{k,n}(t) \geq 0, \quad \forall \, k, \, \forall \, n \qquad (141)$$

Notice that the cost function in (138) is different from the one in (132), since we have dispensed with the expectation and time-averaging in (138). This can be done, because the fractions $w_{k,n}(t)$ at time t, can be optimally chosen *independent of future channel gains*, thanks to the Reference Channel formulation. Thus, interestingly, whereas the optimal scheduler state would need to incorporate the channel states of the $N \times K$ fading channels (if they are correlated between different frames t), our sub-optimal scheduler does not need to do so, since the correlations in the channel fading process have been already accounted for by the admission control algorithm!

It is obvious that our sub-optimal scheduling problem (*i.e.*, the minimization problem (138)) is simply a linear program. The solution (scheduler) can be found with low complexity, by either the simplex method or interior-point methods [97, pp. 362–417].

The constraint (139) is for the case of fixed channel assignment (associated with RR scheduling). If the admission control and resource allocation algorithm in Section 6.2.2.2 is used, the constraint (139) becomes

$$\sum_{n=1}^{N} w_{k,n}(t)c_{k,n}(t) \geq \sum_{n=1}^{N}(\zeta + \beta \times \mathbf{1}(k = k^*(n,t)))c_{k,n}(t), \qquad \forall\, k \qquad (142)$$

where $k^*(n,t)$ is the index of the user whose capacity $c_{k,n}(t)$ is the largest among K users, for channel n, and $\mathbf{1}(\cdot)$ is an indicator function such that $\mathbf{1}(k = a) = 1$ if $k = a$, and $\mathbf{1}(k = a) = 0$ if $k \neq a$. Note that if $\zeta = 0$, *i.e.*, the admission control algorithm allocates channel resources to K&H scheduling only, then the RC scheduler is equivalent to the K&H scheduling. This is because we then have

$$w_{k,n}(t) = \beta \times \mathbf{1}(k = k^*(n,t))), \qquad \forall\, k, \forall n, \qquad (143)$$

which means for each channel, the best user is chosen to transmit, and this is exactly the same as the K&H scheduling. So the relation between the joint K&H/RR scheduling and the RC scheduling is that 1) if the admission control allocates channel resources to the RR scheduling due to tight delay requirements, then the RC scheduler can be used to minimize channel usage; 2) if the admission control allocates channel resources to the K&H scheduling only with $\beta = 1$, due to loose delay requirements, then there is no need to use the RC scheduler. The second statement is formally presented in the following proposition.

Proposition 6.1 *Assume K users share N parallel channels and the K users are scheduled by the K&H scheduling specified in Section 6.2.2.1. If $\beta = 1$, then there does not exist a channel assignment $\{w_{k,n}(t) : k = 1, \cdots, K; n = 1, \cdots, N\}$ such that $\sum_{k=1}^{K}\sum_{n=1}^{N} w_{k,n}(t) < N$.*

154

Table 6: Channel capacities $c_{k,n}(t)$.

	Channel 1	Channel 2
User 1	9	3
User 2	1	5

For a proof of Proposition 6.1, see Appendix A.13. Proposition 6.1 states that if the K users are scheduled by the K&H scheduler with $\beta = 1$, then no channel assignment $\{w_{k,n}(t) : k = 1, \cdots, K; n = 1, \cdots, N\}$ can reduce the channel resource usage.

Next we show a toy example of the capacity gain achieved by the RC scheduler over the RR scheduler. Suppose $K = N = 2$ and the channel capacities at frame t are listed in Table 6. Also assume that channel allocation $\xi_{1,1} = \xi_{1,2} = \xi_{2,1} = \xi_{2,2} = 1/2$ so that the two channels are completely allocated. Then, using the RC scheduler, at frame t, user 1 will be assigned with 2/3 of channel 1 and user 2 will be assigned with 3/5 of channel 2. Hence, the resulting channel usage is $2/3 + 3/5 = 19/15 < 2$. So the channel usage of the RC scheduler is reduced, as compared to the RR scheduler.

In the next section, we investigate the performance of our RC scheduler. In particular, since the optimal scheduler (based on dynamic programming) is very complex, we present a simple bound for evaluating the performance of the RC scheduler. Then, in Section 6.4 we show that the performance of the RC scheduler is close to the bound.

6.3.3 Performance Analysis

To evaluate the performance of the RC scheduling algorithm, we introduce two metrics, expected channel usage $\eta(K, N)$ and expected gain $L(K, N)$ defined as below,

$$\eta(K, N) = \frac{\frac{1}{\tau} \sum_{t=0}^{\tau-1} \mathbf{E}[\sum_{k=1}^{K} \sum_{n=1}^{N} w_{k,n}(t)]}{N}, \tag{144}$$

where the expectation is over $g_{k,n}(t)$, and

$$L(K, N) = \frac{1}{\eta(K, N)} \tag{145}$$

The quantity $1 - \eta(K, N)$ represents average free channel resource (per channel), which can be used for supporting the users, other than the QoS-assured

K users. For example, the frame fractions $\{1 - \sum_k w_{k,n}(t)\}$ of each channel n, which are unused after the K users have been supported, can be used for either Best Effort (BE) or Guaranteed Rate (GR) traffic [49]. It is clear that the smaller the channel usage $\eta(K, N)$ (the larger the gain $L(K, N)$), the more free channel resource is available to support BE or GR traffic. The following proposition shows that minimizing $\eta(K, N)$ or maximizing $L(K, N)$ is equivalent to maximizing the capacity available to support BE/GR traffic.

Proposition 6.2 *Assume that the unused frame fractions $\{1 - \sum_{k=1}^{K} w_{k,n}(t)\}$ are used solely by K_B BE/GR users (indexed by $K + 1, K + 2, \cdots, K + K_B$), whose channel gain processes are i.i.d. (in user k and channel n), strict-sense stationary (in time t) and independent of the K QoS-assured users. If the BE/GR scheduler allots each channel to the contending user with the highest channel gain among the K_B users, then the 'available expected capacity',*

$$C_{exp} = \mathbf{E}\left[\sum_{n=1}^{N}(1 - \sum_{k=1}^{K} w_{k,n}(t))c_{k^*(n,t),n}(t)\right], \qquad (146)$$

is maximized by any scheduler that minimizes $\eta(K, N)$ or maximizes $L(K, N)$. Here, $k^*(n, t)$ denotes the index of the BE/GR user with the highest channel gain among the K_B BE/GR users, for the nth channel in frame t.

For a proof of Proposition 6.2, see Appendix A.14.

Next, we present bounds on $\eta(K, N)$ and $L(K, N)$, which will be used to evaluate the performance of the RC scheduler.

Computing (144) for the optimal scheduler (132) through (135) is complex, because the optimal scheduler itself has high complexity. For this reason, we seek to derive a lower bound on $\eta(K, N)$ of the RC scheduler. We consider the case where K users have i.i.d. channel gains which are stationary processes in frame t. The following proposition specifies a lower bound on $\eta(K, N)$ of the RC scheduler.

Proposition 6.3 *Assume that K users have N i.i.d. channel gains which are strict-sense stationary processes in frame t. Each user k has channel assignments $\{\xi_{k,n}\}$ (for the RR scheduling only), where $\xi_{k,n} = \zeta_k, \forall k, n$. Assume that the N channels are fully assigned to the K users, i.e.,*

$$\sum_{k=1}^{K}\sum_{n=1}^{N} \xi_{k,n} = N \qquad (147)$$

Then a lower bound on $\eta(K, N)$ of the RC scheduler specified by (138) through (141), is

$$\eta(K, N) \geq \mathbf{E}[c_{mean}/c_{max}], \qquad (148)$$

where $c_{mean} = \sum_{n=1}^{N} c_{1,n}/N$ and $c_{max} = \max\{c_{1,1}, c_{1,2}, \cdots, c_{1,N}\}$. The time index has been dropped here, due to the assumption of stationarity of the channel gains. Hence, an upper bound on $L(K, N)$ of the RC scheduler specified by (138) through (141), is

$$L(K, N) \leq \frac{1}{\mathbf{E}[c_{mean}/c_{max}]}. \tag{149}$$

For a proof of Proposition 6.3, see Appendix A.15.

Furthermore, the following proposition states that the upper bound on $L(K, N)$ in (149) monotonically decreases as average SNR increases.

Proposition 6.4 *The lower bound on $\eta(K, N)$ in (148), i.e., $\mathbf{E}[c_{mean}/c_{max}]$, monotonically increases to 1 as SNR_{avg} increases from 0 to ∞, where $SNR_{avg} = P_0/\sigma^2$ [see Eq. (127)]. Hence, the upper bound on $L(K, N)$ in (149), i.e., $1/\mathbf{E}[c_{mean}/c_{max}]$, monotonically decreases to 1 as SNR_{avg} increases from 0 to ∞.*

For a proof of Proposition 6.4, see Appendix A.16. Proposition 6.4 shows that for large SNR_{avg}, there is not much gain to be expected by using the RC scheduler.

So far, we have considered the effect of $\eta(K, N)$ and $L(K, N)$ on the available expected capacity, and derived bounds on $\eta(K, N)$ and $L(K, N)$. In the next section, we evaluate the performance of the RC scheduler and the joint K&H/RR scheduler through simulations.

6.4   Simulation Results

6.4.1   Simulation Setting

We simulate the system depicted in Figure 52, in which each connection[4] is simulated as plotted in Figure 53. In Figure 53, the data source of user k generates packets at a *constant* rate $r_s^{(k)}$. Generated packets are first sent to the (infinite) buffer at the transmitter. The head-of-line packet in the queue is transmitted over N fading channels at data rate $\sum_{n=1}^{N} r_{k,n}(t)$. Each fading channel n has a random power gain $g_{k,n}(t)$. We use a fluid model, that is, the size of a packet is infinitesimal. In practical systems, the results presented here will have to be modified to account for finite packet sizes.

We assume that the transmitter has perfect knowledge of the current channel gains $g_{k,n}(t)$ in frame t. Therefore, it can use rate-adaptive transmissions,

---

[4]Assume that K connections are set up and each mobile user is associated with only one connection.

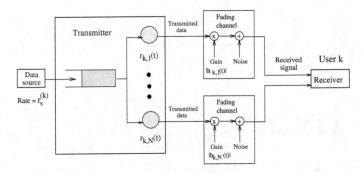

Figure 53: Queueing model used for multiple fading channels.

and ideal channel codes, to transmit packets without decoding errors. Under the joint K&H/RR scheduling of Section 6.2, the transmission rate $r_{k,n}(t)$ of user k over channel n, is given as below,

$$r_{k,n}(t) = (\zeta + \beta \times \mathbf{1}(k = k^*(n,t)))c_{k,n}(t), \tag{150}$$

where the instantaneous channel capacity $c_{k,n}(t)$ is

$$c_{k,n}(t) = B_c \log_2(1 + g_{k,n}(t) \times P_0/\sigma^2) \tag{151}$$

where B_c is the channel bandwidth. On the other hand, for the combination of joint K&H/RR and RC scheduling (Section 6.3), the transmission rate $r_{k,n}(t)$ of user k over channel n, is given as below,

$$r_{k,n}(t) = w_{k,n}(t)c_{k,n}(t). \tag{152}$$

where $\{w_{k,n}(t)\}$ is a solution to the linear program specified by (138), (140), (141) and (142).

The average SNR is fixed in each simulation run. We define r_{awgn} as the capacity of an equivalent AWGN channel, which has the same average SNR. That is,

$$r_{awgn} = B_c \log_2(1 + SNR_{avg}) \tag{153}$$

where $SNR_{avg} = E[g_{k,n}(t) \times P_0/\sigma^2] = P_0/\sigma^2$, assuming that the transmission power P_0 and noise variance σ^2 are constant and equal for all users, in a simulation run. We set $E[g_{k,n}(t)] = 1$ and $r_{awgn} = 1000$ kb/s in all the simulations.

The sample interval (frame length) T_s is set to 1 milli-second and each simulation run is 100-second long in all scenarios. Denote $h_{k,n}(t)$ the voltage gain of the nth channel for the kth user. As described in Section 4.4.1, we generate Rayleigh flat-fading voltage-gains $h_{k,n}(t)$ by a first-order auto-regressive (AR(1)) model as below. We first generate $\bar{h}_{k,n}(t)$ by

$$\bar{h}_{k,n}(t) = \kappa \times \bar{h}_{k,n}(t-1) + u_{k,n}(t), \tag{154}$$

where $u_{k,n}(t)$ are i.i.d. complex Gaussian variables with zero mean and unity variance per dimension. Then, we normalize $\bar{h}_{k,n}(t)$ and obtain $h_{k,n}(t)$ by

$$h_{k,n}(t) = \bar{h}_{k,n}(t) \times \sqrt{\frac{1-\kappa^2}{2}}. \tag{155}$$

In all the simulations, we set $\kappa = 0.8$, which roughly corresponds to a Doppler rate of 58 Hz.

We only consider the homogeneous case, *i.e.*, each user k has the same QoS requirements $\{r_s^{(k)}, \rho_k\}$, and the channel gain processes $\{g_{k,n}(t)\}$ are i.i.d in channel n and user k (note that $g_{k,n}(t)$ is not i.i.d. in t).

6.4.2 Performance Evaluation

We organize this section as follows. In Section 6.4.2.1, we assess the accuracy of our QoS estimation (108). In Section 6.4.2.2, we evaluate the performance of our joint K&H/RR scheduler. In Section 6.4.2.3, we evaluate the performance of our RC scheduler.

6.4.2.1 Accuracy of Channel Estimation

The experiments in this section are to show that the estimated effective capacity can indeed be used to accurately predict QoS.

In the experiments, the following parameters are fixed: $K = 40$, $N = 4$, and $SNR_{avg} = -40$ dB. By changing the source rate μ, we simulate three cases, *i.e.*, μ =200, 300, and 400 kb/s. Figure 54 shows the actual delay-bound violation probability $Pr\{D(\infty) > D_{max}\}$ vs. the delay bound D_{max}. From the figure, it can be observed that the actual delay-bound violation probability decreases exponentially with D_{max}, for all the cases. This confirms the exponential dependence shown in (108). In addition, the estimated $Pr\{D(\infty) > D_{max}\}$ is quite close to the actual $Pr\{D(\infty) > D_{max}\}$, which demonstrates the effectiveness of our channel estimation algorithm.

6.4.2.2 Performance Gain of Joint K&H/RR Scheduling

The experiments here are intended to show the performance gain of the joint K&H/RR scheduler in Section 6.2.2.1 due to utilization of multiple channels.

159

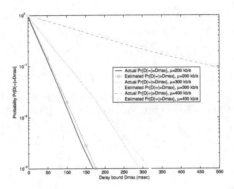

Figure 54: Actual and estimated delay-bound violation probability.

This can be compared with Section 5.4 where only a single channel was assumed.

We set $SNR_{avg} = -40$ dB. The experiments use the optimum $\{\zeta, \beta\}$ values specified by the resource allocation algorithm, *i.e.*, Eqs. (128)–(131). For a fair comparison, we fix the ratio N/K so that each user is allotted the same amount of channel resource for different $\{K, N\}$ pairs. We simulate three cases: 1) $K = 10$, $N = 1$, 2) $K = 20$, $N = 2$, 3) $K = 40$, $N = 4$. For Case 1, the joint K&H/RR scheduler in Section 6.2.2.1 reduces to the joint scheduler presented in Chapter 5.

In Figure 55, we plot the function $\theta(\mu)$ achieved by the joint, K&H, and RR schedulers under Case 3, for a range of source rate μ, when the entire frame of each channel is used (*i.e.*, $K\zeta + \beta = 1$). The function $\theta(\mu)$ in the figure is obtained by the estimation scheme described in [158]. In the case of joint scheduling, each point in the curve of $\theta(\mu)$ corresponds to a specific optimum $\{\zeta, \beta\}$, while $K\zeta = 1$ and $\beta = 1$ are set for RR and K&H scheduling respectively. The curve of $\theta(\mu)$ can be directly used to check for feasibility of a QoS pair $\{r_s, \rho\}$, by checking whether $\theta(r_s) > \rho$ is satisfied. From the figure, we observe that the joint scheduler has a larger effective capacity than both the K&H and the RR for a rather small range of θ. Therefore, in practice, it may be sufficient to use either K&H or RR scheduling, depending on whether θ is small or large respectively, and dispense with the more complicated joint scheduling. Cases 1 and 2 have similar behavior to that plotted in Figure 55.

Figure 56 plots the function $\theta(\mu)$ achieved by the joint K&H/RR scheduler in three cases, for a range of source rate μ, when the entire frame is used (*i.e.*, $K\zeta + \beta = 1$). This figure shows that the larger N is, the higher capacity the

160

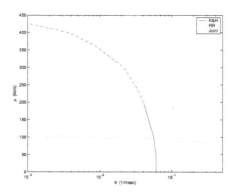

Figure 55: $\theta(\mu)$ vs. μ for K&H, RR, and joint scheduling $(K = 40, N = 4)$.

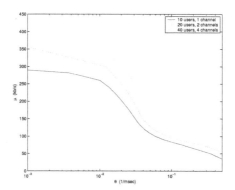

Figure 56: $\theta(\mu)$ vs. μ for joint K&H/RR scheduling.

161

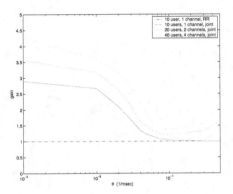

Figure 57: Gain for joint K&H/RR scheduling over RR scheduling.

joint K&H/RR scheduler in Section 6.2.2.1 achieves, given each user allotted the same amount of channel resource. This is because the larger N is, the higher diversity the scheduler can achieve. For small θ, the capacity gain is due to multiuser diversity, *i.e.*, there are more users as N increases for fixed N/K; for large θ, the capacity gain is achieved by frequency diversity, *i.e.*, there are more channels to be simultaneously utilized as N increases.

On the other hand, using the RR scheduler for single channel as a benchmark, we plot the capacity gain achieved by the joint K&H/RR scheduler in Figure 57. The capacity gain of the joint scheduler is the ratio of $\mu(\theta)$ of the joint scheduler to the $\mu(\theta)$ of the RR scheduler. For $N \geq 2$, the figure shows that 1) in the range of small θ, the capacity gain decreases with the increase of θ, which is due to the fact that multiuser diversity is less effective as θ increases, 2) in the range of large θ, the capacity gain increases with the increase of θ, which is due to the fact that the effect of frequency diversity kicks in as θ increases, 3) in the middle range of θ, the capacity gain is the least since both multiuser diversity and frequency diversity are less effective.

The simulation results in this section demonstrate that the joint K&H/RR scheduler can significantly increase the delay-constrained capacity of fading channels, compared with the RR scheduling, for any delay requirement; and the joint K&H/RR scheduler for the multiple channel case achieves higher capacity gain than that for the single channel case.

162

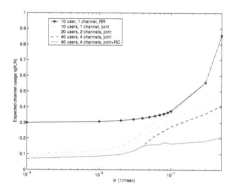

Figure 58: Expected channel usage $\eta(K, N)$ vs. θ.

6.4.2.3 Performance Gain of RC Scheduling

The experiments in this section are aimed to show the performance gain achieved by the RC scheduler.

We simulate three scenarios for the experiments. In the first scenario, we change the QoS requirement θ while fixing other source/channel parameters. We fix the data rate $r_s^{(k)} = 30$ kb/s to compare the difference in channel usage achieved by different schedulers. In this scenario, the N channels are not fully allocated by the admission control. Figure 58 shows the expected channel usage $\eta(K, N)$ vs. θ for the RR scheduler, joint K&H/RR scheduler (denoted by "joint" in the figure), and the combination of joint K&H/RR and the RC scheduler (denoted by "joint+RC" in the figure). It is noted that for $N \geq 2$, the joint K&H/RR scheduler uses less channel resources than the RR scheduler for any θ, and the combination of the joint K&H/RR and the RC scheduler further reduces the channel usage, for large θ. We also observe that 1) for small θ, the K&H scheduler suffices to minimize the channel usage (the RC scheduling does not help since the RC scheduling only improves over the RR scheduling); 2) for large θ, the RC scheduler with fixed channel assignment achieves the minimum channel usage (the K&H scheduler does not help since the K&H scheduler is not applicable for large θ).

In the second and third scenarios, we only simulate the RC scheduler with fixed channel assignment. In the experiments, we choose $\{r_s^{(k)}, \rho_k\}$ so that $\theta_{\zeta,\beta}(r_s^{(k)}) = \rho_k$, where $\zeta = 1/K$ and $\beta = 0$. Hence, the N channels are fully allocated to K users by the admission control, and we have fixed channel assignment $\xi_{k,n} = \zeta$, $\forall k, \forall n$. We set $K = N$ since the performance gain

163

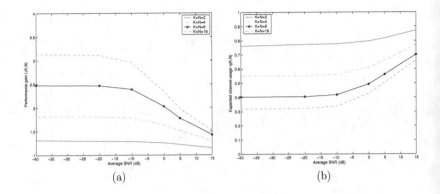

(a)   (b)

Figure 59: (a) Performance gain $L(K, N)$ vs. average SNR, and (b) $\eta(K, N)$ vs. average SNR.

$L(K, N)$ will remain the same for the same N and any $K \geq N$, if the channels are fully allocated to the K users by the admission control.

In the second scenario, we change the average SNR of the channels while fixing other source/channel parameters. Figure 59(a) shows performance gain $L(K, N)$ vs. average SNR. Just as Proposition 6.4 indicates, the gain $L(K, N)$ monotonically decreases as the average SNR increases from –40 dB to 15 dB. Intuitively, this is caused by the concavity of the capacity function $c = \log_2(1 + g)$. For high average SNR, a higher channel gain does not result in a substantially higher capacity. Thus, for a high average SNR, scheduling by choosing the best channels (with or without QoS constraints) does not result in a large $L(K, N)$, unlike the case of low average SNR. In addition, Figure 59(a) shows that the gain $L(K, N)$ falls more rapidly for larger N. This is because a larger N results in a larger $L(K, N)$ at low SNR while at high SNR, $L(K, N)$ goes to 1 no matter what N is (see Proposition 6.4). Figure 59(b) shows the corresponding expected channel usage vs. average SNR.

In the third scenario, we change the number of channels N while fixing other source/channel parameters. Figure 60 shows the performance gain $L(K, N)$ versus number of channels N, for different average SNRs. It also shows the upper bound (149). From the figure, we observe that as the number of channels increases from 2 to 16, the gain $L(K, N)$ increases. This is because a larger number of channels in the system, increases the likelihood of using channels with large gains, which translates into higher performance gain. Another interesting observation is that the performance gain $L(K, N)$ increases almost linearly with the increase of $\log_e N$ (note that the X-axis in the figure is in a log scale). We also plot the corresponding expected channel usage $\eta(K, N)$ vs. number of channels in Figure 61. The lower bound in Figure 61 is computed by (148). One may notice that the gap between the bound and the actual metric in Figs. 60 and 61 reduces as the number of channels increases. This is because the more channels there is, the less the channel usage is, and hence the more likely each user chooses its best channel to transmit, so that the actual performance gets closer to the bound.[5]

For all the simulations, we verify that the actual QoS achieved by the RC scheduler meets the users' requirements. The actual delay-bound violation probability curve is similar to that in Figure 46 and is upper-bounded by the requested delay-bound violation probability. These observations are not shown here.

In summary, the joint K&H/RR scheduler for the multiple channel case achieves higher capacity gain than that for the single channel case; the RC scheduler further squeezes out the capacity from multiple channels, when delay requirements are tight.

6.5 Related Work

There have been many proposals on QoS provisioning in wireless networks. Since our work is centered on scheduling, we will focus on the literature on scheduling with QoS constraints in wireless environments. Besides K&H scheduling and Bettesh & Shamai's scheduler that we discussed in Section 5.1, previous works on this topic also include wireless fair queueing [87, 94, 106], modified largest weighted delay first (M-LWDF) [4], opportunistic transmission scheduling [83] and lazy packet scheduling [102].

Wireless fair queueing schemes [87, 94, 106] are aimed at applying Fair Queueing [99] to wireless networks. The objective of these schemes is to provide fairness, while providing loose QoS guarantees. However, the problem formulation there does not allow explicit QoS guarantees (e.g., explicit delay bound or rate guarantee), unlike our approach. Further, their problem formulation stresses fairness, rather than efficiency, and hence, does not utilize multiuser diversity to improve capacity.

The M-LWDF algorithm [4] and the opportunistic transmission scheduling [83] implicitly utilize multiuser diversity, so that higher efficiency can be achieved. However, the schemes do not provide explicit QoS, but rather optimize a certain QoS parameter.

---

[5]In the proof of Proposition 6.3, we show that the bound corresponds to the case where each user chooses its best channel to transmit.

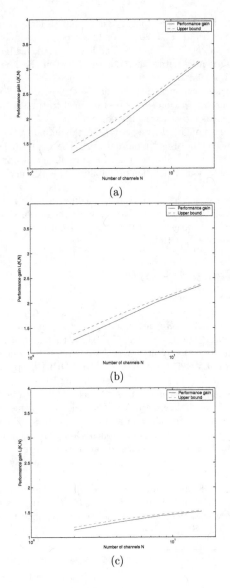

Figure 60: $L(K, N)$ vs. number of channels N for average SNR = (a) –40 dB, (b) 0 dB, and (c) 15 dB.   166

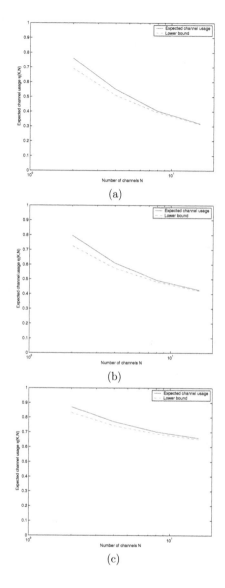

Figure 61: $\eta(K, N)$ vs. number of channels N for average SNR = (a) −40 dB, (b) 0 dB, and (c) 15 dB.

The lazy packet scheduling [102] is targeted at minimizing energy, subject to a delay constraint. The scheme only considers AWGN channels and thus allows for a deterministic delay bound, unlike fading channels and the general statistical QoS considered in our work.

Static fixed channel assignments, primarily in the wireline context, have been considered [61], in a multiuser, multichannel environment. However, these do not consider channel fading, or general QoS guarantees.

Time-division scheduling has been proposed for 3-G WCDMA [60, page 226]. The proposed time-division scheduling is similar to the RR scheduling in this chapter and Chapter 5. However, their proposal did not provide methods on how to use time-division scheduling to support statistical QoS guarantees explicitly. With the notion of effective capacity, we are able to make explicit QoS provisioning with our joint scheduling.

As mentioned in Section 6.3.2, the RC scheduling approach has similarities to the various scheduling algorithms, which use a 'Virtual time reference', such as Virtual Clock, Fair Queueing (and its packetized versions), Earliest Deadline Due, etc. These scheduling algorithms handle source randomness, by prioritizing the user transmissions, using an easily-computed sequence of transmission times. A scheduler that follows the transmission times, is guaranteed to satisfy the QoS requirements of the users. Similarly, in our work, channel randomness is handled by allotting users an easily-computed 'Virtual channel reference', $i.e.$, the channel assignment $\{\xi_{k,n}\}$. A scheduler (of which the RC scheduler is the optimum version) that allots the time-varying capacities specified by $\{\xi_{k,n}\}$, at each time instant, is guaranteed to satisfy the QoS requirements of the users (assuming an appropriate admission control algorithm was used in the calculation of $\{\xi_{k,n}\}$).

6.6 Summary

In this chapter, we examined the problem of providing QoS guarantees to K users over N parallel time-varying channels. We designed simple and efficient admission control, resource allocation, and scheduling algorithms for guaranteeing requested QoS. We developed two sets of scheduling algorithms, namely, joint K&H/RR scheduling and RC scheduling. The joint K&H/RR scheduling utilizes both multiuser diversity and frequency diversity to achieve capacity gain, and is an extension of Chapter 5. The RC scheduling is formulated as a linear program, which minimizes the channel usage while satisfying users' QoS constraints. The relation between the joint K&H/RR scheduling and the RC scheduling is that 1) if the admission control allocates channel resources to the RR scheduling due to tight delay requirements, then the RC scheduler can be used to minimize channel usage; 2) if the admission control allocates channel resources to the K&H scheduling only, due to loose delay requirements, then there is no need to use the RC scheduler.

The key features of the RC scheduler are,

- High efficiency. This is achieved by dynamically selecting the best channel to transmit.

- Simplicity. Dynamic programming is often required to provide an optimal solution to the scheduling problem. However, the high complexity of dynamic programming (exponential in the number of users) prevents it from being used in practical implementations. On the other hand, the RC scheduler has a low complexity (polynomial in the number of users), and yet performs very close to the bound we derived. This indicates that the RC scheduler is simple and efficient.

- Statistical QoS support. The RC scheduler is targeted at statistical QoS support. The statistical QoS requirements are represented by the channel assignments $\{\xi_{k,n}\}$, which appear in the constraints of the linear program of the scheduler.

Simulation results have demonstrated that substantial gain can be achieved by the joint K&H/RR scheduler and the RC scheduler, and have validated our analysis of the RC scheduler performance.

CHAPTER 7

PRACTICAL ASPECTS OF EFFECTIVE

CAPACITY APPROACH

In Chapters 4 to 6, we made some ideal assumptions on the system under consideration such as ergodic, stationary channel gain processes and ideal channel codes. In practical situations, non-ideal conditions could potentially impact the performance of our effective capacity approach. In this chapter, we address the following practical aspects.

- Modulation and channel coding

  In Chapters 4 to 6, we used Shannon's channel capacity to represent the instantaneous maximum error-free data rate, which is typically not achievable by practical modulation and channel coding schemes. Hence, for practical QoS provisioning, we need to consider the effect of modulation and channel coding.

- Robustness against non-stationary channel gain processes

  Our estimation algorithm was developed under the assumption of ergodic, stationary channel gain processes. In practice, the average power may change over time due to changes in the distance between the transmitter and the receiver and due to shadowing, while the Doppler rate may change over time due to changes in the velocity of the source/receiver (*i.e.*, non-stationary small scale fading). Such non-stationary behavior, which is possible in practical channels, can cause severe QoS violations. So, it is important to design QoS provisioning mechanisms that can mitigate large scale fading as well as non-stationary small scale fading, using our channel estimation algorithm.

To familiarize the reader with practical situations, we list typical QoS requirements for different wired/wireless applications in Table 7 [41, 46, 60]; for a carrier frequency of 1.9 GHz, the Doppler rate is between 0 and 350 Hz if the velocity of a mobile terminal is between 0 and 200 km/h.

The rest of this chapter is organized as below. Section 7.1 discusses the effect of modulation and channel coding on the effective capacity approach to

170

Table 7: QoS requirements for different wired/wireless applications.

Class	Application	Data rate	Delay bound	BER
Interactive	Voice	2.4 – 64 kb/s	30 – 60 ms	10^{-3}
real-time	Video-conference	128 kb/s – 6 Mb/s	40 – 90 ms	10^{-3}
Non-interactive	Streaming audio	5 kb/s – 1.4 Mb/s	1 s – 10 minutes	10^{-5}
real-time	Streaming video	10 kb/s – 10 Mb/s	1 s – 10 minutes	10^{-5}
	Web browsing	None	None	10^{-8}
Non-real-time	File transfer	None	None	10^{-8}
	Email	None	None	10^{-8}

QoS provisioning. In Section 7.2, we identify the trade-off between power control and the utilization of time diversity, and propose a novel time-diversity dependent power control scheme to leverage time diversity. Section 7.3 describes our power control and scheduling scheme, which is robust in QoS provisioning against large scale fading and non-stationary small scale fading. In Section 7.4, we show simulation results to demonstrate the effectiveness of our power control and scheduling scheme under various practical situations. Section 7.5 summarizes the chapter.

7.1 Effect of Modulation and Channel Coding

In Chapters 4 to 6, we used Shannon's channel capacity (52) to represent the instantaneous maximum error-free data rate, which is typically not achievable by practical modulation and channel coding schemes. For practical modulation and channel coding schemes, the instantaneous data rate $r(t)$, which results in negligible decoding error probability (as compared to the target P_{loss}, an upper bound on $Pr\{D(\infty) \geq D_{max}\}$), would have to be used. For example, for quadrature amplitude modulation (QAM), the rate $r(t)$ as a function of the received SNR can be given by

$$r(t) = B_c \log_2 \left(1 + \frac{P_0 \times g(t)}{\Gamma_{link} \times (\sigma_n^2 + P_I(t))} \right), \quad (156)$$

where Γ_{link} is determined by the target decoding error probability and the channel code used, $g(t)$ is the channel power gain at frame t, σ_n^2 is the noise variance and $P_I(t)$ is the instantaneous interference power.

To illustrate how to obtain Γ_{link}, we use M-ary QAM as the modulation scheme, where M is the number of symbols used in the QAM. The channel code can be any practical code, *e.g.*, Reed-Solomon codes, convolutional codes, Turbo codes, or low density parity check (LDPC) codes. For generality purpose, we do not specify the channel code to be used but only assume the coding gain γ_{code} is given. For M-ary QAM and a specified channel code, the symbol error probability P_e is upper-bounded as [103, page 280]

$$P_e \leq 4 \times Q \left(\sqrt{\frac{3 \times P_0 \times g(t) \times \gamma_{code}}{(M-1) \times (\sigma_n^2 + P_I(t))}} \right), \tag{157}$$

where the Q function is given by

$$Q(x) = \frac{1}{\sqrt{2\pi}} \int_x^\infty e^{-x^2/2} \mathrm{d}x. \tag{158}$$

Let

$$\beta_{mod} = \frac{3 \times P_0 \times g(t) \times \gamma_{code}}{(M-1) \times (\sigma_n^2 + P_I(t))}. \tag{159}$$

Denote P_{loss} the packet loss probability due to buffer overflow at the transmitter. Let $4Q(\sqrt{\beta_{mod}})$ be equal to a value, denoted by ϵ_{error}, which is much smaller than P_{loss}. So, the dominant impairment in the transmission is caused by buffer overflow, while the symbol error is negligible. For a given ϵ_{error}, from $4Q(\sqrt{\beta_{mod}}) = \epsilon_{error}$, we can obtain β_{mod} by

$$\beta_{mod} = \left(Q^{-1}(\epsilon_{error}/4) \right)^2, \tag{160}$$

where $Q^{-1}(\cdot)$ is the inverse of the Q function. Representing M by 2^b, we get

$$b = \log_2 \left(1 + \frac{3 \times P_0 \times g(t) \times \gamma_{code}}{\beta_{mod} \times (\sigma_n^2 + P_I(t))} \right) \qquad \text{bits/s/Hz}. \tag{161}$$

Hence, the rate $r(t)$ is given by

$$r(t) = B_c \log_2 \left(1 + \frac{3 \times P_0 \times g(t) \times \gamma_{code}}{\beta_{mod} \times (\sigma_n^2 + P_I(t))} \right). \tag{162}$$

So, in this case, Γ_{link} is given by

$$\Gamma_{link} = \beta_{mod}/(3\gamma_{code}). \tag{163}$$

Similarly, we can obtain the rate $r(t)$ as a function of the received SNR for phase shift keying (PSK). For M-ary PSK, the rate $r(t)$ can be given by

$$r(t) = B_c \log_2 \left(\frac{\pi}{\arcsin \left(\beta_{mod} \times \sqrt{\frac{(\sigma_n^2 + P_I(t))}{2 \times P_0 \times g(t) \times \gamma_{code}}} \right)} \right), \qquad (164)$$

where

$$\beta_{mod} = Q^{-1}(\epsilon_{error}/2). \qquad (165)$$

To summarize, for practical modulation and channel coding schemes, we use the instantaneous data rate $r(t)$, which results in negligible decoding error probability. Generally speaking, the rate $r(t)$ is some function of the received SNR, which could differ from the specific form as (52). With this $r(t)$, we can apply the effective capacity approach developed in Chapters 4 to 6.

7.2 Trade-off between Power Control and Time Diversity

It is well known that ideal power control can completely eliminate fading and convert the fading channel to an AWGN channel, so that deterministic QoS (zero queueing delay and zero delay-bound violation probability) can be guaranteed. However, fast fading (time diversity) is actually useful. From the link-layer perspective, as mentioned in Section 4.3.3, the higher the degree of time diversity, the larger the QoS exponent $\theta(\mu)$ for a fixed μ, and the larger the effective capacity $\alpha(u)$ for a fixed u. But for a slow fading channel, we know that the effective capacity $\alpha(u)$ can be very small due to the stringent delay requirement, and therefore power control may be needed to provide the required QoS. Hence, it is conceivable that there is a trade-off between power control and the utilization of time diversity, depending on the degree of time diversity and the QoS requirements.

To identify this trade-off, we compare the following three schemes through simulations:

- **Ideal power control:** In order to keep the received SINR constant at a target value $SINR_{target}$, the transmit power at frame t is determined as below

$$P_0(t) = \frac{SINR_{target}}{\tilde{g}(t)}, \qquad (166)$$

where the channel power gain $\tilde{g}(t)$ (absorbing the noise variance plus interference) is given by

$$\tilde{g}(t) = \frac{g(t)}{\sigma_n^2 + P_I(t)} \tag{167}$$

Denote P_{avg} the time average of $P_0(t)$ specified by (166); the time average is over the entire simulation duration. Note that the fast power control used in 3G networks is an approximation of ideal power control, in that the fast power control in 3G has a peak power constraint and in that the power change (in dB) in each interval can only be a fixed integer, say 1 dB, rather than an arbitrary real number as in (166).

- **Fixed power:** The transmit power $P_0(t)$ is kept constant and is equal to P_{avg}. The objective of this scheme is to use time diversity only.

- **Time-diversity dependent power control:** To utilize time diversity, the transmit power at frame t is determined as below

$$P_0(t) = \frac{\gamma_{coeff}}{g_{avg}(t)}, \tag{168}$$

where γ_{coeff} is so determined that the time average of $P_0(t)$ in (168) is equal to P_{avg}; and $g_{avg}(t)$ is given by an exponential smoothing of $\tilde{g}(t)$ as below

$$g_{avg}(t) = (1 - \eta_g) \times g_{avg}(t-1) + \eta_g \times \tilde{g}(t) \tag{169}$$

where $\eta_g \in [0,1]$ is a fixed parameter, chosen depending on the time diversity desired. It is clear that if $\eta_g = 0$, the time-diversity dependent power control reduces to the fixed power scheme; if $\eta_g = 1$, the time-diversity dependent power control reduces to ideal power control. Hence, by optimally selecting $\eta_g \in [0,1]$, we expect to trade-off time diversity against power control.

The three schemes have been so specified that they use the same amount of average power P_{avg}, for fairness of comparison. In all of the three schemes, the transmission rate at frame t is given as

$$r(t) = B_c \log_2 \left(1 + \frac{P_0(t) \times \tilde{g}(t)}{\Gamma_{link}} \right), \tag{170}$$

with the assumption that $\tilde{g}(t)$ is perfectly known at the transmitter.

From the simulations in Section 7.4.2.2 and Figure 62, we have the following observations:

174

1. *Power control vs. using time diversity.* If the degree of time diversity is low, ideal power control provides a substantial capacity gain as opposed to the fixed power scheme, which only uses time diversity; otherwise, schemes utilizing time diversity can provide a higher effective capacity than ideal power control. The reason is as follows. When the degree of time diversity is low, which implies that the probability of having long deep fades is high, then ideal power control can keep the error-free data rate $r(t)$ constant at a high value even during deep fades, while the fixed power scheme suffers from low data rate $r(t)$ during deep fades. On the other hand, when the degree of time diversity is high and hence the probability of having long deep fades is small, one can leverage time diversity by buffering data during deep fades (limited by the delay bound D_{max}) and transmitting at a high data rate when the channel conditions are good.

2. The maximum data rate μ, with $Pr\{D(\infty) > D_{max}\} \leq \varepsilon$ satisfied, denoted by $\hat{\mu}(D_{max}, \varepsilon)$, under both the fixed power control and the time-diversity dependent power control, increases with the degree of time diversity. The reason is as given above.

3. The time-diversity dependent power control, which jointly utilizes power control and time diversity, is the best among the three schemes. This is because the fixed power scheme and ideal power control are special cases of the time-diversity dependent power control, when $\eta_g = 0$ and 1, respectively. Hence, by optimally selecting $\eta_g \in [0, 1]$, the time-diversity dependent power control can achieve the largest $\hat{\mu}(D_{max}, \varepsilon)$.

4. As the degree of time diversity increases, the capacity gain provided by the time-diversity dependent power control increases as compared to ideal power control; the capacity gain provided by the time-diversity dependent power control decreases as compared to the fixed power scheme. This is because, as the degree of time diversity increases, the effect of time diversity on $\hat{\mu}(D_{max}, \varepsilon)$ increases, while the effect of power control on $\hat{\mu}(D_{max}, \varepsilon)$ does not change.

In this section, we identified the tradeoff between power control and the utilization of time diversity, and designed a novel scheme called *time-diversity dependent power control*, which achieves the largest data rate $\hat{\mu}(D_{max}, \varepsilon)$, among the three schemes. In the next section, we present QoS provisioning algorithms that utilize time-diversity dependent power control.

7.3 Power Control and Scheduling

In this section, we present schemes to achieve robustness in QoS provisioning against large scale path loss, shadowing, non-stationary small scale fading, and very low mobility. The basic elements of the schemes are channel estimation, power control, dynamic channel allocation, and adaptive transmission.

7.3.1 Downlink Transmission

We first describe the schemes for the case of downlink transmissions, *i.e.*, a base station (BS) transmits data to a mobile station (MS).

Assume that a connection requesting a QoS triplet $\{r_s, D_{max}, \varepsilon\}$ or equivalently $\{r_s, \rho = -\log \varepsilon / D_{max}\}$, is accepted by the admission control (described later in Algorithm 7.2). In the transmission phase, the following tasks are performed.

1. **SINR estimation at the MS:** The MS estimates instantaneous received SINR at frame t, denoted by $SINR(t)$, which is given by

$$SINR(t) = \frac{P_0(t) \times g(t)}{\sigma_n^2 + P_I(t)}. \tag{171}$$

   Then the MS conveys the value of $SINR(t)$ to the BS. Since the value of $SINR(t)$ is typically within a range of 30 dB, *e.g.*, from -19 to 10 dB, five bits should be enough to represent the value of $SINR(t)$ to within 1 dB. If the estimation frequency is 200 Hz, the signaling overhead is only 1 kb/s, which is low. Note that 3G allows for a 1500-Hz power control loop.

2. **Time-diversity dependent power control at the BS:** Since the BS knows the transmit power $P_0(t)$, upon receiving $SINR(t)$, it can derive the channel power gain $\tilde{g}(t)$ as below

$$\tilde{g}(t) = \frac{SINR(t)}{P_0(t)} = \frac{g(t)}{\sigma_n^2 + P_I(t)} \tag{172}$$

   Denote P_{peak} the peak transmit power at the BS. The BS determines the transmit power for frame $t + 1$ by

$$P_0(t+1) = \min \left\{ \frac{SINR_{target}}{g_{avg}(t)}, P_{peak} \right\}, \tag{173}$$

   where $g_{avg}(t)$ is given by (169). Note that η_g in (169) is time-diversity dependent; based on the current mobile speed $v_s(t)$, the value of η_g is specified by a table, similar to Table 9.

176

Note that the downlink power control described here is different from the downlink power control in 3G systems, in that in our scheme, the BS initiates the power control while in 3G systems, the MS initiates the power control. Specifically, in our system, the BS determines the transmit power 'value' based on the value of $SINR(t)$ sent by the MS, while in 3G systems, the power control signal (*i.e.*, power-up or power-down signal) is sent from the MS to the BS. In 3G systems, the power-up signal requests an increase of transmit power by a preset value, *e.g.*, 1 dB, and the power-down signal requests a decrease of transmit power by a preset value, *e.g.*, 1 dB.

3. **Estimation of QoS exponent θ at the BS:** The BS measures the queueing delay $D(t)$ at the transmit buffer, and estimates the average queueing delay $D_{avg}(t)$ at frame t by

$$D_{avg}(t) = (1 - \eta_d) \times D_{avg}(t - 1) + \eta_d \times D(t) \qquad (174)$$

where $\eta_d \in (0, 1)$ is a preset constant. Then, the BS estimates the QoS exponent θ at frame t, denoted by $\hat{\theta}(t)$, as below

$$\hat{\theta}(t) = \frac{1}{0.5 + D_{avg}(t)} \qquad (175)$$

Note that the estimation in Eq. (175) is different from the estimation in Eqs. (36) through (39), in that the expected queueing delay in (35) is approximated by the exponentially smoothed $D_{avg}(t)$ in (175), while the expected queueing delay in (35) is approximated by the time-averaged queueing delay in (39).[1]

4. **Scheduling (dynamic channel allocation) at the BS:**

From Figure 62, we know that as the degree of time diversity, or equivalently the mobile speed, increases (resp., decreases), the data rate $\hat{\mu}(D_{max}, \varepsilon)$ increases (resp., decreases) and the QoS exponent $\theta(\mu = r_s)$ increases (resp., decreases), hence requiring less (resp., more) channel resource to support the QoS. This motivates us to design a dynamic channel allocation mechanism that can adapt to changes in channel statistics, to achieve both efficiency and QoS guarantees.

The basic idea of dynamic channel allocation is to use the QoS measures $\hat{\theta}(t)$ and $D(t)$ in deciding channel allocation. Specifically, the BS

---

[1] The expected queueing delay is equal to the expected queue length divided by the source rate.

allocates a fraction $\lambda(t+1)$ of frame $t+1$, to the connection, as below

$$
\lambda(t+1) = \begin{cases} \min\{\lambda(t)+\Delta_\lambda, 1\} & \text{if } \hat{\theta}(t) < \gamma_{inc} \times \rho \text{ and } D(t) > D_h; \\ \max\{\lambda(t)-\Delta_\lambda, 0\} & \text{if } \hat{\theta}(t) > \gamma_{dec} \times \rho \text{ and } D(t) < D_l; \\ \lambda(t) & \text{otherwise.} \end{cases}
$$
(176)

where $\Delta_\lambda \in (0,1)$, $\gamma_{inc} \geq 1$, $\gamma_{dec} \geq \gamma_{inc}$, low threshold $D_l \in (0, D_{max})$, and high threshold $D_h \in (D_l, D_{max})$ are preset constants.

It is clear that the control in (176) has hysteresis (due to $D_h > D_l$ and $\gamma_{dec} \geq \gamma_{inc}$), which helps reduce the variation in $\lambda(t)$ and hence reduce the signaling overhead for dynamic channel allocation. The condition $\hat{\theta} < \gamma_{inc} \times \rho$ means that the measured QoS exponent $\hat{\theta}$ does not meet the required ρ, scaled by $\gamma_{inc} \geq 1$ to allow a safety margin; the condition $D(t) > D_h$ means that the delay $D(t)$ is larger than the high threshold D_h; the two conditions jointly trigger an increase in $\lambda(t)$. Similarly, the condition $\{\hat{\theta} > \gamma_{dec} \times \rho \text{ and } D(t) < D_l\}$ causes a decrease in $\lambda(t)$.

In practice, $\lambda(t)$ can be interpreted in different ways, depending on the type of the system. For CDMA, TDMA, and FDMA systems, $\lambda(t)$ can be implemented by using variable spreading codes, variable number of mini-slots, and variable number of frequency carriers, respectively.

For ease of implementation, one can set $\Delta_\lambda = 0.1$ so that $\lambda(t)$ only takes discrete values from the set $\{0, 0.1, 0.2, \cdots, 0.9, 1\}$. Then, in a TDMA system, if a frame consists of ten mini-slots, $\lambda(t) = 0.3$ would mean using three mini-slots to transmit the data at frame t; the remaining seven mini-slots in the frame can be used by other users, *e.g.*, best-effort users.

5. **Adaptive transmission at the BS:** Once the channel allocation $\lambda(t+1)$ is given, the BS determines the transmission rate at frame $t+1$ as below

$$
\begin{aligned}
r(t+1) &= \lambda(t+1) \times \varpi^* \times B_c \times \log_2\left(1 + \frac{P_0(t+1) \times g(t)}{(\sigma_n^2 + P_I(t)) \times \Gamma_{link} \times \gamma_{safe}}\right) \quad (177) \\
&= \lambda(t+1) \times \varpi^* \times B_c \times \log_2\left(1 + \frac{P_0(t+1) \times SINR(t)}{P_0(t) \times \Gamma_{link} \times \gamma_{safe}}\right) \quad (178)
\end{aligned}
$$

where ϖ^* denotes the amount of channel resource allocated by the admission control (described later in Algorithm 7.2), Γ_{link} characterizes the effect of practical modulation and coding and γ_{safe} introduces a safety margin to mitigate the effect of the SINR estimation error at the MS. The BS uses (178) to compute $r(t+1)$ since all variables in (178) are known.

The values of Γ_{link} and γ_{safe} are so chosen that transmitting at the rate $r(t+1)$ specified by (178) will result in negligible bit error rate (w.r.t. P_{loss}). So, $r(t+1)$ specified by (178) can be regarded as an error-free data rate. Since (178) takes into account the effect of the physical layer (*i.e.*, practical modulation, channel coding, and SINR estimation error), we can focus on the queueing behavior and link-layer performance.

Once $r(t+1)$ is determined, an M-ary QAM can be used for the transmission, where $M = 2^b$ and b is given by

$$b = floor\left(\log_2\left(\lambda(t+1) \times \varpi^* \times \log_2\left(1 + \frac{P_0(t+1) \times SINR(t)}{P_0(t) \times \Gamma_{link} \times \gamma_{safe}}\right)\right)\right)$$
(179)

where $floor(x)$ is the largest integer that is not larger than x.

The above tasks are summarized in Algorithm 7.1.

Algorithm 7.1 Downlink power control, channel allocation, and adaptive transmission

In the transmission phase, the following tasks are performed.

1. **SINR estimation at the MS:** *The MS estimates the received $SINR(t)$ and conveys the value of $SINR(t)$ to the BS.*

2. **Power control at the BS:** *The BS derives the channel power gain $\tilde{g}(t)$ using (172), estimates $g_{avg}(t)$ using (169), and then determines the transmit power $P_0(t+1)$ using (173).*

3. **Estimation of QoS exponent θ at the BS:** *The BS measures the queueing delay $D(t)$, estimates $D_{avg}(t)$ using (174), and estimates the QoS exponent $\hat{\theta}(t)$ using (175).*

4. **Scheduling at the BS:** *The BS allocates a fraction of frame $\lambda(t+1)$ to the connection, using (176).*

5. **Adaptive transmission at the BS:** *The BS determines the transmission rate $r(t+1)$ using (178).*

The key elements in Algorithm 7.1 are power control and scheduling. The power control is intended to mitigate large scale path loss, shadowing, and low mobility. The scheduler specified by (176) is targeted at achieving both efficiency and QoS guarantees.

In Algorithm 7.1, the power control allocates the power resource, while the scheduler allocates the channel resource; their effects on the 'error-free' transmission rate $r(t)$ in (178) are different: $r(t)$ is linear in channel allocation $\lambda(t)$, but is a log-function of power $P_0(t)$.

Remark 7.1 Power control vs. channel allocation in QoS provisioning

From (178), we see that the error-free data rate $r(t)$ is determined by the channel resource allocated $\lambda(t)$ and power $P_0(t)$. A natural question is how to optimally allocate the channel and power resource to satisfy the required QoS.

There are two extreme cases. First, if the transmit power $P_0(t)$ is fixed and we suppose $\lambda(t) \in [0, \infty)$, then given arbitrary channel gain $g(t)$ (which includes the effect of the noise and interference), we can obtain arbitrary $r(t) \in [0, \infty)$ by choosing appropriate $\lambda(t) \in [0, \infty)$. Second, if the channel resource allocated $\lambda(t)$ is fixed and we suppose $P_0(t) \in [0, \infty)$, then given arbitrary channel gain $g(t)$, we can obtain arbitrary $r(t) \in [0, \infty)$ by choosing appropriate $P_0(t) \in [0, \infty)$.

However, in practical situations, we have both a peak power constraint $P_0(t) \leq P_{peak}$ and a peak channel usage constraint $\lambda(t) \leq 1$, assuming that $\lambda(t)$ is the fraction of allotted channel resource. Hence, we cannot obtain arbitrary $r(t) \in [0, \infty)$, given arbitrary channel gain $g(t)$. Since applications can tolerate a certain delay and there is a buffer at the link layer, $r(t)$ is allowed to be less than the arrival rate, with a small probability. Therefore, there could be feasible solutions $\{P_0(t), \lambda(t)\}$ that satisfy the QoS constraint, peak power constraint, and peak channel usage constraint. If such feasible solutions do exist, the next question is which one is the optimal solution, given a certain criterion. If we want to minimize average power usage (resp., average channel usage) under the QoS constraint, peak power constraint, and peak channel usage constraint, an optimal solution must have $\lambda(t) = 1$ (resp., $P_0(t) = P_{peak}$). Hence, we cannot simultaneously minimize both average power or average channel usage; and we are facing a multi-objective optimization problem. A classical multi-objective optimization method is to convert a multi-objective optimization problem to a single-objective optimization problem by a weighted sum of multiple objectives, the solution of which is Pareto optimal [36, page 49]. Using this method, we *formulate an optimization problem as follows*

$$\underset{\{P_0(t),\lambda(t):t=0,1,\cdots,\tau-1\}}{maximize} \quad \frac{1}{\tau}\sum_{t=0}^{\tau-1}\mathbf{E}[\beta_{weight}\times P_0(t)+(1-\beta_{weight})\times\lambda(t)] \quad (180)$$

$$subject\ to \quad Pr\{D(\infty)\geq D_{max}\}\leq\varepsilon, \qquad for\ a\ fixed\ rate\ r_s \ (181)$$

$$0\leq P_0(t)\leq P_{peak} \qquad\qquad\qquad\qquad (182)$$

$$0\leq\lambda(t)\leq 1 \qquad\qquad\qquad\qquad (183)$$

where τ is the connection life time, and $\beta_{weight}\in[0,1]$. Dynamic programming often turns out to be a natural way to solve (180). However, the complexity of solving the dynamic program is high. If the statistics of the channel gain process are unpredictable (due to large scale fading and time-varying mobile speed), we cannot use dynamic programming to solve (180). This motivates us to seek a simple (sub-optimal) approach, which can enforce the specified QoS constraints explicitly, and yet achieve an efficient channel and power usage. Our scheme is based on the tradeoff between power and time diversity: we use the time-diversity dependent power control to maximize the data rate $\hat{\mu}(D_{max},\varepsilon)$, and use scheduling to determine the minimum amount of resource that satisfies the required QoS, given the choice of power control. This leads to two separate optimization problems, which simplifies the complexity, while achieving good performance. Algorithm 7.1 is designed according to this idea.

Now, we get to the issue of admission control. Assume that a user initiates a connection request, requiring a QoS triplet $\{r_s, D_{max},\varepsilon\}$. In the connection setup phase, we use Algorithm 7.2 (see below) to test whether the required QoS can be satisfied. Specifically, the algorithm measures the QoS that the link-layer channel can provide; if the measured QoS satisfies the required QoS, the connection request is accepted; otherwise, it is rejected.

Algorithm 7.2 uses the methods in Algorithm 7.1. The key difference between the two algorithms is that in Algorithm 7.2, the BS creates a fictitious queue, that is, the BS uses r_s as the arrival rate and $r(t)$ as the service rate to 'simulate' a fictitious queue, but no actual packet is transmitted over the wireless channel. In the admission test, there is no need for the BS to transmit actual data in order to obtain QoS measures of the link-layer channel. This is because 1) the MS can use the common pilot channel [60, page 103] to measure the received $SINR(t)$, and 2) the simulated fictitious queue provides the same queueing behavior as if actual data was transmitted over the wireless channel.

To facilitate resource allocation, we simulate N_{fic} fictitious queues, each of which is allocated with different amount of resource ϖ_i $(i=1,\cdots,N_{fic})$. Assume that ϖ_i represents the proportion of the resource allocated to queue i, to the total resource, and ϖ_i $(i=1,\cdots,N_{fic})$ takes a discrete value in $(0,1]$. If the connection is accepted, the BS allocates the minimum amount of resource (denoted by ϖ^*) that satisfies the QoS requirements, to the connection. That is, ϖ^* is the minimum of all feasible ϖ_i that satisfy the QoS requirements. The algorithm for admission control and resource allocation is as below.

Algorithm 7.2 Downlink admission control and resource allocation: *Upon the receipt of a connection request requiring a QoS triplet $\{r_s, D_{max}, \varepsilon\}$, the following tasks are performed.*

1. **SINR estimation at the MS:** *The MS estimates the received $SINR(t)$ using the common pilot channel, and conveys the value of $SINR(t)$ to the BS.*

2. **Power control at the BS:** *The BS derives the channel power gain $\tilde{g}(t)$ using (172), where $P_0(t)$ is meant to be the actual transmit power for the common pilot channel at frame t. Then, the BS estimates $g_{avg}(t)$ by computing (169). Finally, the BS determines the fictitious transmit power $\tilde{P}_0(t+1)$ using (173).*

3. **Estimation of QoS exponent θ at the BS:** *For each fictitious queue i $(i = 1, \cdots, N_{fic})$, the BS generates fictitious arrivals with data rate r_s, measures the queueing delay $D_i(t)$, estimates $D_{avg}^{(i)}(t)$ using (174), and estimates the QoS exponent $\hat{\theta}_i(t)$ using (175).*

4. **Scheduling at the BS:** *For each fictitious queue i $(i = 1, \cdots, N_{fic})$, the BS allocates a fraction of frame $\lambda_i(t+1)$, using (176).*

5. **Adaptive transmission at the BS:** *For each fictitious queue i $(i = 1, \cdots, N_{fic})$, the BS determines the transmission rate $r_i(t+1)$ as below*

$$r_i(t+1) = \lambda_i(t+1) \times \varpi_i \times B_c \times \log_2 \left(1 + \frac{\tilde{P}_0(t+1) \times SINR(t)}{\tilde{P}_0(t) \times \Gamma_{link} \times \gamma_{safe}} \right) \tag{184}$$

6. **Admission control and resource allocation:** *If there exists a queue $\tilde{i}$ such that its QoS exponent average $\hat{\theta}_{avg}^{(i)}(t) = \frac{1}{t+1} \sum_{\tau=0}^{t} \hat{\theta}_{\tilde{i}}(\tau)$ is not less than a preset threshold θ_{th}, accept the connection request; otherwise, reject it. If the connection is accepted, the BS allocates the minimum amount of resource $\varpi^* = \min_{\tilde{i}} \varpi_{\tilde{i}}$, to the connection.*

Note that in Algorithm 7.2, the MS needs to convey $SINR(t)$ to the BS in the connection setup phase, which is different from the current 3G standard.

182

It is required that Algorithm 7.2 be fast and accurate in order to implement it in practice. Our simulation results in Section 7.4.2.8 show that $\hat{\theta}_{avg}(t)$ is a reliable QoS measure for the purpose of admission control; moreover, within a short period of time, say two seconds, the system can obtain a reasonably accurate $\hat{\theta}_{avg}(t)$ and hence can make a quick and accurate admission decision.

As long as the large scale path loss and shadowing can be mitigated by the power control in (173), the required QoS can be guaranteed. It is known that the large scale path loss within a coverage area can be mitigated by the power control. To mitigate shadowing more effectively as compared to power control, our scheme can be improved by macro-diversity, which employs the collaboration of multiple base stations. We leave this for future study.

7.3.2 Uplink Transmission

For uplink transmissions, *i.e.*, an MS transmits data to a BS, the design methodology for QoS provisioning is the same as that for downlink transmissions. Specifically, we use Algorithms 7.3 and 7.4, which are modifications of Algorithms 7.2 and 7.1. Algorithms 7.3 uses the common random access channel [60, page 106] instead of the common pilot channel as in Algorithm 7.2.

Algorithm 7.3 Uplink admission control and resource allocation:
Upon the receipt of a connection request requiring a QoS triplet $\{r_s, D_{max}, \varepsilon\}$, the following tasks are performed.

1. **SINR estimation at the BS:** *The MS transmits a signal of constant power P_{MS} over the common random access channel to the BS. The value of P_{MS} is known to the BS. The BS estimates received $SINR(t)$ from the common random access channel.*

2. **Power control at the BS:** *The BS derives the channel power gain $\tilde{g}(t)$ by (172), where $P_0(t)$ is equal to P_{MS}. Then the BS estimates $g_{avg}(t)$ by computing (169). Finally, the BS determines the fictitious transmit power $\tilde{P}_0(t+1)$ by (173), where P_{peak} is with respect to the MS and is specified in the 3G standard.*

3. **Estimation of QoS exponent θ at the BS:** *For each fictitious queue i $(i = 1, \cdots, N_{fic})$, the BS generates fictitious arrivals with data rate r_s, measures the queueing delay $D_i(t)$, estimates $D_{avg}^{(i)}(t)$ using (174), and estimates the QoS exponent $\hat{\theta}_i(t)$ using (175).*

4. **Scheduling at the BS:** *For each fictitious queue i $(i = 1, \cdots, N_{fic})$, the BS allocates a fraction of frame $\lambda_i(t+1)$, using (176).*

183

5. **Adaptive transmission at the BS:** *For each fictitious queue i ($i = 1, \cdots, N_{fic}$), the BS determines the transmission rate $r_i(t + 1)$ using (184).*

6. **Admission control and resource allocation:** *If there exists a queue $\tilde{i}$ such that its QoS exponent average $\hat{\theta}_{avg}^{(i)}(t) = \frac{1}{t+1} \sum_{\tau=0}^{t} \hat{\theta}_{\tilde{i}}(\tau)$ is not less than a preset threshold θ_{th}, accept the connection request; otherwise, reject it. If the connection is accepted, the BS allocates the minimum amount of resource $\varpi^* = \min_{\tilde{i}} \varpi_{\tilde{i}}$, to the connection.*

Algorithm 7.4 Uplink power control, channel allocation, and adaptive transmission

In the transmission phase, the following tasks are performed.

1. **SINR estimation at the BS:** *The BS estimates received $SINR(t)$ and conveys the value of $SINR(t)$ to the MS.*

2. **Power control at the MS:** *The MS derives the channel power gain $\tilde{g}(t)$ by (172) and estimates $g_{avg}(t)$ by computing (169). Then, the MS determines the transmit power $P_0(t + 1)$ by (173).*

3. **Estimation of QoS exponent θ at the MS:** *The MS measures the queueing delay $D(t)$, estimates $D_{avg}(t)$ by (174), and estimates the QoS exponent $\hat{\theta}(t)$ using (175).*

4. **Renegotiation of channel allocation:** *The MS computes $\lambda(t + 1)$, using (176). The MS sends a renegotiation request to the BS, asking for a fraction of frame $\lambda(t + 1)$ for the connection. Based on the resource availability, the BS determines the value of $\lambda(t + 1)$, and then notifies the MS of the final value of $\lambda(t + 1)$, which will be used by the MS in frame $t + 1$.*

5. **Adaptive transmission at the MS:** *The MS determines the transmission rate $r(t + 1)$ by (178).*

7.3.3 Multiuser Case

Sections 7.3.1 and 7.3.2 address the case of a single user. In this section, we consider the case of multiple users. Our objective is to extend the results in Chapters 5 and 6. So, we address two cases as below.

Case 1: multiple users sharing a single channel

We consider downlink transmission only and assume the same architecture as that in Chapter 5. Assume that K users have different QoS requirements

184

and are classified into M classes, each of which has the same QoS pair $\{r_s, \rho\}$. For example, voice applications typically have similar QoS pair $\{r_s, \rho\}$. We list some typical traffic classes in Table 7.

We would like to apply the joint K&H/RR scheduling in Section 5.3.2 to each of the M classes. Assume that for each class m, channel allocation $\{\zeta_m, \beta_m\}$ is determined by the resource allocation algorithm (refer to Section 5.3.2). To schedule the K users, we only need to modify the adaptive transmission (178) in Algorithm 7.1 as follows.

For user k in class m, the BS determines the transmission rate at frame $t + 1$ as below

$$r_k(t+1) = \lambda_k(t+1) \times \varpi_m^*(t) \times B_c \times \log_2 \left(1 + \frac{P_0^{(k)}(t+1) \times SINR_k(t)}{P_0^{(k)}(t) \times \Gamma_{link} \times \gamma_{safe}} \right) \tag{185}$$

where the index k denotes user k. The value of $\varpi_m^*(t)$ is determined as follows: if user k has the largest channel gain $\tilde{g}_k(t)$ among all the users in class m, then we set $\varpi_m^*(t) = \zeta_m + \beta_m$; otherwise, we set $\varpi_m^*(t) = \zeta_m$. It is clear that (185) reflects the joint K&H/RR scheduling in Section 5.3.2.

Case 2: multiple users sharing multiple channels

We consider downlink transmission only and assume the same architecture as that in Chapter 6, $i.e.$, K users requiring the same QoS $\{r_s, D_{max}, \varepsilon\}$ or equivalently $\{r_s, \rho\}$, share N independent channels.

We would like to apply the joint K&H/RR scheduling in Section 6.2.2 to the K users. Assume that channel allocation $\{\zeta, \beta\}$ is determined by the resource allocation algorithm (refer to Section 6.2.2). To schedule the K users, we only need to modify the adaptive transmission (178) in Algorithm 7.1 as follows.

For each user k ($k = 1, \cdots, K$) and each channel n ($n = 1, \cdots, N$), the BS determines the transmission rate at frame $t + 1$ as below

$$r_{k,n}(t+1) = \lambda_{k,n}(t+1) \times \varpi_{k,n}^*(t) \times B_c \times \log_2 \left(1 + \frac{P_0^{(k,n)}(t+1) \times SINR_{k,n}(t)}{P_0^{(k,n)}(t) \times \Gamma_{link} \times \gamma_{safe}} \right) \tag{186}$$

where the index k and n denote user k and channel n, respectively. The value of $\varpi_{k,n}^*$ is determined as follows: for channel n, if user k has the largest channel gain $\tilde{g}_{k,n}(t)$ among all the users, then we set $\varpi_{k,n}^* = \zeta + \beta$; otherwise, we set $\varpi_{k,n}^* = \zeta$. It is clear that (186) reflects the joint K&H/RR scheduling in Section 6.2.2.

If the RC scheduler is used, then for each user k and each channel n, the BS determines the transmission rate $r_{k,n}(t+1)$ at frame $t+1$, using (186) with $\varpi_{k,n}^* = w_{k,n}^*$, where $\{w_{k,n}^*\}$ is the optimal solution of (138).

7.4 Simulation Results

In this section, we simulate the discrete-time wireless communication system as depicted in Figure 26, and demonstrate the performance of our algorithms. We focus on Algorithm 7.1 for downlink transmission of a single connection, since the performance of Algorithm 7.4 for uplink transmission would be the same as that for Algorithm 7.1 if the simulation parameters are the same and fast feedback of channel gains is assumed. Section 7.4.1 describes the simulation setting, while Section 7.4.2 illustrates the performance of our algorithms.

7.4.1 Simulation Setting

7.4.1.1 Mobility Pattern Generation

We simulate the speed behavior of the MS using the model described in Ref. [11]. Under the model, an MS moves away from the BS, at a constant speed v_s for a random duration; then a new target speed v^* is randomly generated; the MS linearly accelerates or decelerates until this new speed v^* is reached; following which, the MS moves at the constant speed v^*, and the procedure repeats again.

The speed behavior of an MS at frame t can be described by three parameters:

- its current speed $v_s(t) \in [0, v_{max}]$ in units of m/s

- its current acceleration $a_s(t) \in [a_{min}, a_{max}]$ in m/s^2

- its current target speed $v^*(t) \in [0, v_{max}]$

where v_{max} denotes the maximum speed, a_{min} the minimum acceleration (which is negative), and a_{max} the maximum acceleration.

At the beginning of the simulation, the MS is assigned an initial speed $v_s(0)$, which is generated by a probability density function $f_v(v_s)$, given by

$$f_v(v_s) = \begin{cases} p_0 \times \delta(v_s) & \text{if } v_s = 0; \\ p_{max} \times \delta(v_s - v_{max}) & \text{if } v_s = v_{max}; \\ \frac{1 - p_0 - p_{max}}{v_{max}} & \text{if } 0 < v_s < v_{max}; \\ 0 & \text{otherwise.} \end{cases} \qquad (187)$$

where $p_0 + p_{max} < 1$. That is, the random speed has high probabilities at speed 0 (imitating stops due to red lights or traffic jams) and at the maximum speed

186

v_{max} (a preferred speed when driving); and it is uniformly distributed between 0 and v_{max}.

The speed change events are modeled as a Poisson process. That is, the time between two consecutive speed changes is exponentially distributed with mean m_{v^*}. Note that a speed change event happens at an epoch determined by the Poisson process but it does not include the speed changes during acceleration/deceleration periods.

Now, we know the epochs of speed change events follow a Poisson process and the new target speed v^* follows the PDF $f_v(v_s)$. Denote t^* the time at which a speed change event occurs and $v^* = v^*(t^*)$ the associated new target speed. Then, an acceleration $a_s(t^*) \neq 0$ is generated by the PDF

$$f_a(a_s) = \begin{cases} \frac{1}{a_{max}} & \text{if } 0 < a_s \leq a_{max}; \\ 0 & \text{otherwise.} \end{cases} \quad (188)$$

if $v^*(t^*) > v_s(t^*)$, or by the PDF

$$f_a(a_s) = \begin{cases} \frac{1}{|a_{min}|} & \text{if } a_{min} \leq a_s < 0; \\ 0 & \text{otherwise.} \end{cases} \quad (189)$$

if $v^*(t^*) < v_s(t^*)$. Obviously, a_s is set to 0 if $v^*(t^*) = v_s(t^*)$. If $a_s(t) \neq 0$, the speed continuously increases or decreases; at frame t, a new speed $v_s(t)$ is computed according to

$$v_s(t) = v_s(t-1) + a_s(t) \times T_s \quad (190)$$

until $v_s(t)$ reaches $v^*(t^*)$; T_s is the frame length in units of second. Then, we set $a_s = 0$ and the MS moves at constant speed $v_s(t) = v^*(t^*)$ until the next speed change event occurs. Figure 64 shows a trace of the speed behavior of an MS.

7.4.1.2   Channel Gain Process Generation

The channel power gain process $g(t)$ is given by

$$g(t) = g_{small}(t) \times g_{large}(t) \times g_{shadow}(t) \quad (191)$$

where $g_{small}(t)$, $g_{large}(t)$, and $g_{shadow}(t)$ denote channel power gains due to small-scale fading, large scale path loss, and shadowing, respectively.

Non-stationary small scale fading

Given the mobile speed $v_s(t)$, the Doppler rate $f_m(t)$ can be calculated by [109, page 141]

$$f_m(t) = v_s(t) \times \cos\varphi \times f_c/c, \qquad (192)$$

where φ is the angle between the direction of motion of the MS and the direction of arrival of the electromagnetic waves, f_c is the carrier frequency and c is the speed of light, which is 3×10^8 m/sec. We choose $\varphi = 0$ in all the simulations.

We assume Rayleigh flat-fading for the small scale fading. As described in Section 4.4.1, Rayleigh flat-fading voltage-gains $h(t)$ are generated by an AR(1) model as below. We first generate $\bar{h}(t)$ by

$$\bar{h}(t) = \kappa(t) \times \bar{h}(t-1) + u_g(t), \qquad (193)$$

where $u_g(t)$ are i.i.d. complex Gaussian variables with zero mean and unity variance per dimension. Then, we normalize $\bar{h}(t)$ and obtain $h(t)$ by

$$h(t) = \bar{h}(t) \times \sqrt{\frac{1 - [\kappa(t)]^2}{2}}. \qquad (194)$$

$\kappa(t)$ is determined by 1) computing the Doppler rate $f_m(t)$ for given mobile speed $v_s(t)$, using (192), 2) computing the coherence time T_c by (57), and 3) calculating (58). Then we obtain $g_{small}(t) = |h(t)|^2$.

Large scale path loss

Next, we describe the generation of large scale path loss. Denote $\{x_t, y_t, z_t\}$ and $\{x_r, y_r, z_r\}$ the 3-dimensional locations of the transmit antenna and the receive antenna, respectively. Specifically, z_t and z_r are the heights of the transmit antenna and the receive antenna, respectively. The initial distance d_0 between the MS and BS is given by

$$d_0 = \sqrt{(x_t - x_r)^2 + (y_t - y_r)^2}. \qquad (195)$$

Denote $d_{tr}(t)$ the distance between the BS (transmitter) and the MS (receiver) at t. Hence, we have $d_{tr}(0) = d_0$. Assume that the MS moves directly away from the BS. Then, for $t > 0$, we have

$$d_{tr}(t) = d_{tr}(t-1) + v_s(t) \times T_s. \qquad (196)$$

We use two path loss models: Friis free space model and the ground reflection model. Friis free space model is given by [109, page 70]

$$g_{large}(t) = \left(\frac{c}{f_c \times 4 \times \pi \times d_{tr}(t)} \right)^2, \qquad (197)$$

where c is light speed, and f_c is carrier frequency. The ground reflection (two-ray) model [109, page 89] is given as below

$$g_{large}(t) = \frac{z_t^2 \times z_r^2}{[d_{tr}(t)]^4}. \tag{198}$$

We need to compute the cross-over distance d_{cross} to determine which model to use. d_{cross} is given by [109, page 89]

$$d_{cross} = \frac{20 \times \pi \times z_t \times z_r \times f_c}{3 \times c}. \tag{199}$$

If $d_{tr}(t) \leq d_{cross}$, we choose Friis free space model (197) to generate $g_{large}(t)$; otherwise, we use the ground reflection model (198) to generate $g_{large}(t)$.

Shadowing

We generate the shadow fading process $\tilde{g}_{shadow}(t)$ in units of dB by an AR(1) model as below [53]

$$\tilde{g}_{shadow}(t) = \kappa_{shadow}^{v_s(t) \times T_s / D_{shadow}} \times \tilde{g}_{shadow}(t-1) + \sigma_{shadow} \times \tilde{u}_g(t) \tag{200}$$

where κ_{shadow} is the correlation between two locations separated by a fixed distance D_{shadow}, $\tilde{u}_g(t)$ are i.i.d. Gaussian variables with zero mean and unity variance, σ_{shadow} is a constant in units of dB, $v_s(t)$ is obtained from the above mobility pattern generation, and hence $v_s(t) \times T_s$ is the distance that the MS traverses in frame t. It is obvious that the shadowing gain $g_{shadow}(t) = 10^{\tilde{g}_{shadow}(t)/10}$ follows a log-normal distribution with standard deviation σ_{shadow}.

7.4.1.3 Simulation Parameters

Table 8 lists the parameters used in our simulations. Since we target at interactive real-time applications, we set the QoS triplet as below: $r_s = 50$ kb/s, $D_{max} = 50$ msec, and $\varepsilon = 10^{-3}$. In addition, we set the values of γ_{inc}, γ_{dec}, D_l, and D_h in (176) in such a way that can reduce the signaling overhead for dynamic channel allocation, while meeting the QoS requirements $\{r_s, \rho = -\log_e \varepsilon / D_{max}\}$. Further, we set the values of σ_{shadow}, κ_{shadow}, and D_{shadow} according to Ref. [53]. The maximum speed $v_{max} = 15.6$ m/s corresponds to 35 miles per hour. We set $P_{peak} = 24$ dBm according to the specification of 3G systems on mobile stations [60, page 159], so that our results are also applicable to uplink transmissions. We assume total intra-cell and inter-cell interference $P_I(t)$ is constant over time.

Assume that the random errors in estimating $SINR(t)$ are i.i.d. Gaussian variables with zero mean and variance σ_{est}^2. Denote the random estimation

Table 8: Simulation parameters.

QoS requirement	Constant bit rate r_s	50 kb/s
	Delay bound D_{max}	50 msec
	Delay-bound violation probability ε	10^{-3}
Channel	Bandwidth B_c	300 kHz
	Sampling-interval (frame length) T_s	1 msec
	Noise plus interference power $\sigma_n^2 + P_I(t)$	-100 dBm
Mobility pattern	Maximum speed v_{max}	15.6 m/s
	Minimum acceleration a_{min}	-4 m/s^2
	Maximum acceleration a_{max}	2.5 m/s^2
	Probability p_0	0.3
	Probability p_{max}	0.3
	Mean time between speed change m_{v^*}	25 sec
Shadowing	Standard deviation σ_{shadow}	7.5 dB
	Correlation κ_{shadow}	0.82
	Distance D_{shadow}	100 m
Receive antenna	x_r	100 m
	y_r	100 m
	Height z_r	1.5 m
Transmit antenna	x_t	0
	y_t	0
	Height z_t	50 m
BS	Carrier frequency f_c	1.9 GHz
	Channel coding gain γ_{code}	3 dB
	Target bit error rate ϵ_{error}	10^{-6}
	Smoothing weight for average delay η_d	0.0005
	Standard deviation of estimation error σ_{est}	1 dB
	Safety margin γ_{safe}	1 dB
Power control	Peak transmission power P_{peak}	24 dBm
	$SINR_{target}$	5 dB
Scheduling	Step size Δ_λ	0.1
	γ_{inc}	1
	γ_{dec}	1
	Low threshold D_l	$0.1 \times D_{max}$
	High threshold D_h	$0.5 \times D_{max}$

error in dB by $\tilde{g}_{est}(t)$. Then, the estimated $SINR(t)$ is given by

$$SINR(t) = \frac{P_0(t) \times g(t) \times 10^{\tilde{g}_{est}(t)/10}}{\sigma_n^2 + P_I(t)} \tag{201}$$

To be realistic, the power $P_0(t+1)$ specified in (173) only takes integer values in units of dB and can only change 1 dB in each frame except in Section 7.4.2.2.

We assume that M-ary QAM is used for modulation. Hence, the transmit data rate $r(t)$ is given by (160) and (162), where σ_n^2 is replaced by $\sigma_n^2 + P_I(t)$.

Each simulation run is 100-second long.

7.4.2 Performance Evaluation

We organize this section as follows. In Section 7.4.2.1, we investigate the effect of modulation and channel coding. Section 7.4.2.2 identifies the trade-off between power control and time diversity. In Section 7.4.2.3, we show the accuracy of the exponentially smoothed estimate of θ. Sections 7.4.2.4 to 7.4.2.7 evaluates the performance of Algorithm 7.1 under four cases, namely, a time-varying mobile speed, large scale path loss, shadowing, and very low mobility. In Section 7.4.2.8, we investigate whether our admission control test in Algorithms 7.2 and 7.3 is quick and accurate.

7.4.2.1 Effect of Modulation and Channel Coding

This experiment is to show the effect of modulation and channel coding.

In the simulation, we use M-ary QAM as the modulation scheme, and assume that certain channel code is employed and it provides a channel coding gain $\gamma_{code} = 3$ dB. Our objective is to achieve the symbol error probability not greater than $\epsilon_{error} = 10^{-6}$. After computing (160) and (163), we obtain $\Gamma_{link} = 6.3$ dB. Hence, we have 6.3 dB loss due to practical modulation and channel coding, as compared to the channel capacity. Then, given $B_c = 300$ kHz, $P_0 = 20$ dBm, $g(t) = -120$ dB, and $\sigma_n^2 = -100$ dBm, the error-free data rate $r(t)$ in (156) is 91 kb/s, while the channel capacity in (52) is 300 kb/s. This shows that the data rate is reduced by a factor of 3.3 due to the practical modulation and channel coding.

7.4.2.2 Power Control vs. Time Diversity

This experiment is to identify the trade-off between power control and time diversity.

We compare the three schemes, namely, ideal power control, the fixed power scheme, and the time-diversity dependent power control, defined in Section 7.2.

Table 9: Simulation parameters.

Speed v_s (m/s)	0.011	0.11	0.23	0.57	1.1	5.7	11	56
Speed v_s (km/h)	0.041	0.41	0.81	2	4.1	20	41	204
Doppler rate f_m (Hz)	0.072	0.72	1.4	3.6	7.2	36	72	358
Coherence time T_c (s)	2.5	0.25	0.125	0.05	0.025	0.005	0.0025	0.0005
D_{max}/T_c	0.02	0.2	0.4	1	2	10	20	100
η_g	0.2	0.08	0.08	0.04	0.04	0.02	0.02	0

All the three schemes use the same amount of average power P_{avg}, for the purpose of fair comparison. Assuming that the channel power gain $\tilde{g}(t)$ is perfectly known by the transmitter, all the three schemes determine the transmission rate at frame t using (170).

In each simulation, we generate Rayleigh fading with fixed mobile speed v_s specified in Table 9. We do not simulate large scale path loss and shadowing. For the time-diversity dependent power control, the smoothing factor η_g in (169) is given by Table 9. Note that for different mobile speeds v_s in Table 9, we use different η_g; the value of η_g is chosen so as to maximize the data rate $\hat{\mu}(D_{max}, \varepsilon)$.

Table 9 lists the parameters used in our simulations, where $D_{max} = 50$ ms. The range of speed v_s is from 0.011 to 56 m/s, which covers both downtown and highway speeds. Doppler rate f_m is computed from v_s using (192) and coherence time T_c is computed from f_m by (57). There is no need to simulate the case for $v_s = 0$ since for $v_s = 0$, the theory gives $\hat{\mu}(D_{max}, \varepsilon) = 200, 0$, and 200 kb/s under ideal power control, the fixed power scheme, and the time-diversity dependent power control, respectively. For $v_s = 0$, we have $\eta_g = 1$ and hence the time-diversity dependent power control reduces to ideal power control.

Figure 62 shows data rate $\hat{\mu}(D_{max}, \varepsilon)$ vs. time diversity index D_{max}/T_c for the three schemes. It is clear that the larger the index D_{max}/T_c is, the higher degree of time diversity the channel possesses.

From the figure, we have the following observations besides those mentioned in Section 7.2:

- Under ideal power control, the data rate $\hat{\mu}(D_{max}, \varepsilon)$ is independent of the degree of time diversity, since ideal power control converts a fading channel to an AWGN channel.

- If the time diversity index D_{max}/T_c is less than a threshold (its value

192

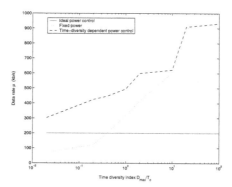

Figure 62: Data rate $\hat{\mu}(D_{max}, \varepsilon)$ vs. time diversity index D_{max}/T_c.

is 10 in Figure 62), the data rate $\hat{\mu}(D_{max}, \varepsilon)$ under the time-diversity dependent power control is larger than that under the fixed power control. This is because when the degree of time diversity is not high, and hence the probability of having long deep fades is not very small, power control (with appropriate smoothing factor η_g) can increase the error-free instantaneous data rate $r(t)$ during deep fades, resulting in higher $\hat{\mu}(D_{max}, \varepsilon)$. On the other hand, if D_{max}/T_c is greater than the threshold, the time-diversity dependent power control and the fixed power control achieve almost the same rate for $\hat{\mu}(D_{max}, \varepsilon)$. This is because there is enough time diversity and the probability of having long deep fades is very small, and hence power control does not give much gain.

- The two curves of ideal power control and the fixed power control intersect at $D_{max}/T_c = 0.4$, *i.e.*, $T_c = D_{max}/0.4 = 125$ ms. This indicates that if $D_{max}/T_c > 0.4$, time diversity takes effect and we should not use ideal power control; otherwise, ideal power control can achieve a larger $\hat{\mu}(D_{max}, \varepsilon)$ as compared to the fixed power control.

- Compared to ideal power control, for speed $v_s = 5.7$ m/s, both the time-diversity dependent power control and the fixed power scheme produce a factor of 3 capacity gain; for $v_s = 56$ m/s, both the time-diversity dependent power control and the fixed power scheme provide a factor of 4.7 capacity gain.

- Compared to ideal power control, the time-diversity dependent power control can achieve a factor of 2 to 4.7 capacity gain for mobile speeds between 0.11 m/s and 56 m/s.

7.4.2.3 Accuracy of Exponentially Smoothed Estimate of θ

This experiment is to show the accuracy of the exponentially smoothed estimate of θ via (174) and (175).

We do experiments with three constant mobile speeds, *i.e.*, $v_s = 5$, 10, and 15 m/s, respectively. For each mobile speed, we do simulations under three values of smoothing factor η_d in (174), *i.e.*, $\eta_d = 5 \times 10^{-4}$, 10^{-4}, and 5×10^{-5}, respectively. Since the objective is to test the accuracy of the estimator of θ, we do not use power control and scheduling; that is, we keep both the power and the channel allocation constant during the simulations.

We set the following parameters: source data rate $r_s = 50$ kb/s, $B_c = 300$ kHz, $P_0 = 20$ dBm, $E[g(t)] = -120$ dB, $\sigma_n^2 + P_I(t) = -100$ dBm, and $\Gamma_{link} = 6.3$ dB. Assume that the transmitter has a perfect knowledge about the channel power gain $g(t)$. The error-free data rate $r(t)$ is calculated by (156).

Figure 63 shows the estimate $\hat{\theta}(t)$ vs. time t under different mobile speed v_s and different smoothing factor η_d. The reference θ in the figure is obtained by computing Eqs. (36) through (39) at $t = 10^5$. It can be observed that for $\eta_d = 5 \times 10^{-5}$, the estimate $\hat{\theta}(t)$ gives the best agreement with the reference θ, as compared to other values of η_d; for $\eta_d = 5 \times 10^{-4}$, the estimate $\hat{\theta}(t)$ reaches the reference θ in the shortest time (within 2 seconds), as compared to other values of η_d.

Hence, for the admission control in Algorithms 7.2 and 7.3, which requires quick estimate of θ, we suggest to use $\eta_d = 5 \times 10^{-4}$; the estimate takes less than 2 seconds, which is tolerable in practice. For Algorithms 7.1 and 7.4, we also suggest to use $\eta_d = 5 \times 10^{-4}$ since we want the estimate $\hat{\theta}(t)$ to be more adaptive to time-varying mobile speed $v_s(t)$ and the resulting queueing behavior.

7.4.2.4 Performance under a Time-varying Mobile Speed

This experiment is to evaluate the performance of Algorithm 7.1 under a time-varying mobile speed, *i.e.*, under non-stationary small scale fading. Our objective is to see whether the power control and the scheduler in Algorithm 7.1 can achieve the required QoS. Note that the efficiency of our time-diversity dependent power control has been addressed in Section 7.4.2.2.

For the channel gain process $g(t)$, we only simulate small scale fading; that is, there are no large scale path loss and shadowing. We set $E[g(t)] = -100$ dB. Other parameters are listed in Table 8.

Figure 64 shows the speed $v_s(t)$ of an MS vs. time t, which is the mobility pattern used in the simulation. Figure 65 depicts the delay $D(t)$ vs. time t. The simulation gives zero delay-bound violation for $D_{max} = 50$ ms, and hence

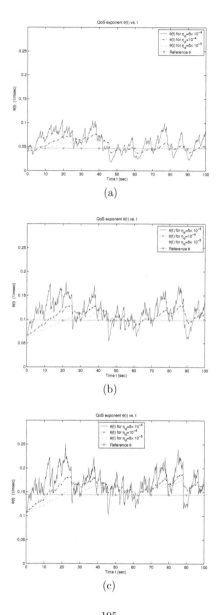

(a)

(b)

(c)

Figure 63: $\hat{\theta}(t)$ vs. time t for speed $v_s =$ (a) 5 m/s, (b) 10 m/s, and (c) 15 m/s.

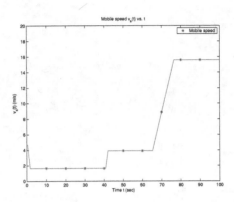

Figure 64: Speed behavior $v_s(t)$ of a mobile station in downtown.

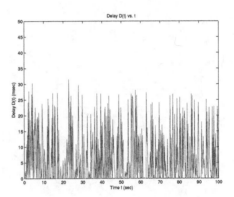

Figure 65: Delay $D(t)$ vs. time t.

196

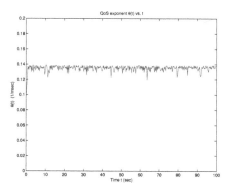

Figure 66: $\hat{\theta}(t)$ vs. time t for varying mobile speed.

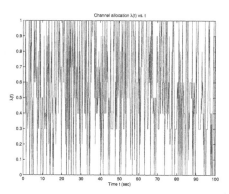

Figure 67: Channel allocation $\lambda(t)$ vs. time t.

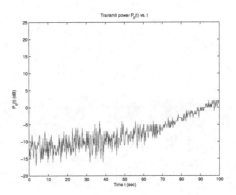

Figure 68: Transmit power $P_0(t)$ vs. time t for large scale path loss.

the required QoS is met. Figure 66 plots QoS exponent $\hat{\theta}(t)$ vs. time t. Since we set $\gamma_{inc} = 1$ and $\gamma_{dec} = 1$, the resulting QoS exponent $\hat{\theta}(t)$ fluctuates around the required $\rho = -\log_e \varepsilon / D_{max} = 0.1382$. Figures 65 and 66 demonstrates the effectiveness of our scheduler in utilizing QoS exponent $\hat{\theta}(t)$ and the delay $D(t)$ for QoS provisioning.

Figure 67 illustrates how the channel allocation $\lambda(t)$ varies with time t. The average channel usage is 0.54.

In summary, Algorithm 7.1 can achieve the required QoS, under non-stationary small scale fading.

7.4.2.5 Performance under Large Scale Path Loss

This experiment is to evaluate the performance of Algorithm 7.1 under large scale path loss. We would like to see how the scheduler and the power control coordinate under large scale path loss.

In the simulation, we use the same mobility pattern as shown in Figure 64 and generate large scale path loss according to Section 7.4.1.2. We do not simulate the shadowing effect here, which will be addressed in Section 7.4.2.6. The simulation parameters are listed in Table 8.

Figure 68 shows how the transmit power $P_0(t)$ evolves over time. The average transmit power is -7.4 dB. The power control is fast with a frequency of 1000 Hz, so that it can utilize time diversity. It is observed that as time elapses, the distance between the transmitter and the receiver increases and hence the expectation of the transmit power increases in order to mitigate the path loss.

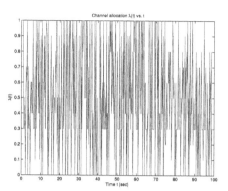

Figure 69: Channel allocation $\lambda(t)$ vs. time t for large scale path loss.

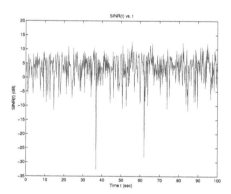

Figure 70: $SINR(t)$ vs. time t for large scale path loss.

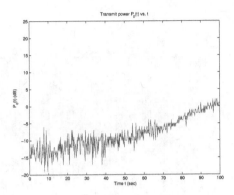

Figure 71: Transmit power $P_0(t)$ vs. time t for AR(1) shadowing.

Figure 69 depicts how the scheduler allocates the channel resource $\lambda(t)$ over time. The simulation gives zero delay-bound violation for $D_{max} = 50$ ms, and hence the required QoS is met. This demonstrates the good coordination between the power control and the scheduler; that is, the power control mitigates large scale path loss, while the scheduler utilizes time diversity in QoS provisioning. The average channel usage is 0.5. Figure 70 plots the received $SINR(t)$ vs. time t.

In summary, we observe the concerted efforts of the scheduler and the power control for QoS provisioning; the power control handles the effects of large scale path loss, while both the power control and the scheduler utilize time diversity. Different from ideal power control, our power control does not eliminate small scale fading, so that time diversity in small scale fading can be utilized.

7.4.2.6   *Performance under Shadowing*

This experiment is to evaluate the performance of Algorithm 7.1 under shadowing.

In the first simulation, we use the same mobility pattern as shown in Figure 64 and generate large scale path loss and AR(1) shadowing process according to Section 7.4.1.2. The simulation parameters are listed in Table 8. Figure 71 depicts the transmit power $P_0(t)$ vs. time t. The simulation gives zero delay-bound violation for $D_{max} = 50$ ms, and hence the required QoS is met. This demonstrates that the power control can mitigate both large scale path loss and shadowing effectively for QoS provisioning.

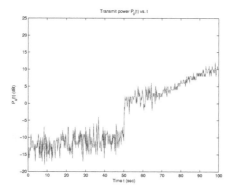

Figure 72: Transmit power $P_0(t)$ vs. time t for the case of sudden shadowing.

In the second simulation, we intentionally generate a shadowing of -10 dB at the 50-th second (which may happen when a car suddenly moves into the 'shadow' of a building) and see whether our power control can adapt and mitigate the shadowing effect. We use the same mobility pattern as shown in Figure 64 and generate large scale path loss according to Section 7.4.1.2. Figure 72 depicts the transmit power $P_0(t)$ vs. time t. It is observed that the power can quickly adapt to the sudden power change caused by the shadowing at the 50-th second in the figure. The simulation gives zero delay-bound violation for $D_{max} = 50$ ms, and hence the required QoS is met. Therefore, our time-diversity dependent power control can also mitigate the sudden shadowing effect.

In summary, Algorithm 7.1 is able to achieve good performance under shadowing.

7.4.2.7 Performance under Very Low Mobility

This experiment is to evaluate the performance of Algorithm 7.1 under very low mobility, especially when the mobile speed is zero (due to red lights or traffic jams). Since our effective capacity approach and the scheduler require time diversity, they are not applicable to the case where the mobile speed is zero. Note that the effective capacity is zero when the mobile speed is zero. Hence, we rely on the power control to provide the required QoS.

Figure 73 shows the mobility pattern used in the simulation. We generate large scale path loss but do not simulate the shadowing effect. We assume perfect estimation of $SINR(t)$. The simulation parameters are listed in Table 8.

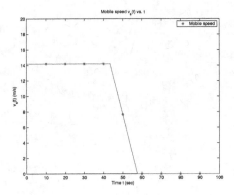

Figure 73: Speed behavior $v_s(t)$ of a mobile station in downtown.

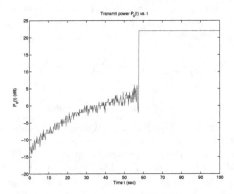

Figure 74: Transmit power $P_0(t)$ vs. time t for the case of very low mobility.

202

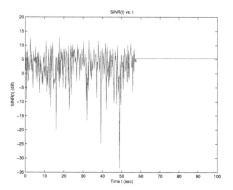

Figure 75: $SINR(t)$ vs. time t for the case of very low mobility.

Figure 74 shows how the transmit power $P_0(t)$ varies over time. Figure 75 plots the received $SINR(t)$ vs. time t; this demonstrates that the power control converts the channel to an AWGN channel when the speed is zero between 57-th second and 100-th second. The simulation gives zero delay-bound violation for $D_{max} = 50$ ms, and hence the required QoS is met.

In summary, our power control can mitigate the effect of very low mobility and Algorithm 7.1 is able to guarantee the required QoS.

7.4.2.8 Admission Control

This experiment is to investigate whether our admission control in Algorithms 7.2 and 7.3 can be done quickly and accurately.

We use previous results in Sections 7.4.2.4 to 7.4.2.7. Define QoS exponent average $\hat{\theta}_{avg}(t) = \frac{1}{t+1} \sum_{\tau=0}^{t} \hat{\theta}(\tau)$. Figure 76 plots $\hat{\theta}_{avg}(t)$ vs. t for the four cases, namely, a time-varying mobile speed, large scale path loss, shadowing with the AR(1) model, and very low mobility, which we investigated in Sections 7.4.2.4 to 7.4.2.7. We set the threshold $\theta_{th} = 0.9 \times \rho$. Since we are only concerned with the quickness of the estimation, we only plot the first ten seconds of the simulations. Figure 76 shows that $\hat{\theta}_{avg}(t)$ is roughly an increasing function of t. Hence, $\hat{\theta}_{avg}(t)$ is a reliable QoS measure for admission control purpose. Moreover, the figure shows that for all the four cases, the system can obtain a reasonably accurate $\hat{\theta}_{avg}(t) \geq \theta_{th}$ within two seconds. Therefore, the system can make a quick and accurate admission decision.

203

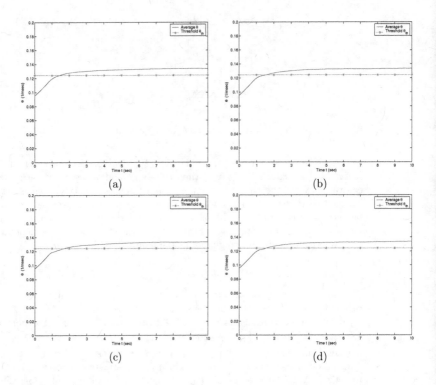

Figure 76: QoS exponent average $\hat{\theta}_{avg}(t)$ vs. t for (a) time-varying mobile speed, (b) large scale path loss, (c) shadowing, and (d) very low mobility.

Table 10: Simulation parameters.

Channel	Noise plus interference power $\sigma_n^2 + P_I(t)$	-100 dBm
	Path loss	100 dB
BS	Carrier frequency f_c	1.9 GHz
	Channel coding gain γ_{code}	3 dB
	Target bit error rate ϵ_{error}	10^{-5}
Power control	Peak transmission power P_{peak}	24 dBm
	$SINR_{target}$	12.5 dB

7.4.2.9   An Example for W-CDMA systems

This experiment is to show an example of applying the K&H/RR scheduler to W-CDMA systems [60], and compare this scheduler to the QoS provisioning schemes used in 3G systems, *i.e.*, ideal power control with the round robin scheduler.

A cellular wireless network is assumed, and the downlink is considered, where a base station transmits data to 5 voice mobile terminals and 5 streaming video mobile terminals. The voice users have the same QoS requirements: constant data rate $r_s = 12.2$ kb/s, delay bound $D_{max} = 50$ msec, and delay-bound violation probability $\varepsilon = 10^{-3}$. The streaming video users have the same QoS requirements: $r_s = 144$ kb/s, $D_{max} = 500$ msec, and $\varepsilon = 10^{-3}$. We set the required BER for all the ten users to be not greater than 10^{-5} and hence the required SNR per bit is 9.5 dB [103, page 270]. Since QPSK is specified for downlink modulation in 3G W-CDMA systems [60, page 85], the required SNR per symbol is 12.5 dB. Assume all users have i.i.d. channel gain processes and the channel gain processes are Rayleigh flat-fading. As described in Section 4.4.1, Rayleigh flat-fading voltage-gains $h_k(t)$ of user k are generated by an AR(1) model as below. We first generate $\bar{h}_k(t)$ by

$$\bar{h}_k(t) = \kappa \times \bar{h}_k(t-1) + u_g^{(k)}(t), \qquad (202)$$

where $u_g^{(k)}(t)$ are i.i.d. complex Gaussian variables with zero mean and unity variance per dimension. Then, we normalize $\bar{h}_k(t)$ and obtain $h_k(t)$ by

$$h_k(t) = \bar{h}_k(t) \times \sqrt{\frac{1-\kappa^2}{2}}. \qquad (203)$$

κ is determined by the velocity of the mobiles $v = 15$ m/s, assuming that all the mobiles have the same velocity.

The chip rate of 3G W-CDMA systems is 3.84 MHz [60]. Hence, the processing gain γ_{proc} is $3.84 \times 10^6/r_s$. The frame size is set to be 1 ms. Table 10 lists the parameters used in this experiment.

The mobile terminals need to convey the values of $SINR(t)$ to the base station for adaptive transmission. Since the value of $SINR(t)$ is typically within a range of 30 dB, $e.g.$, from -19 to 10 dB, five bits should be enough to represent the value of $SINR(t)$ to within 1 dB. Since the estimation frequency in this experiment is 1000 Hz, the signaling overhead is 5 kb/s. Furthermore, in this experiment, the time-diversity dependent power control reduces to the fixed power scheme since the channels have high degree of time diversity.

We do two sets of simulations. In the first simulation, we simulate ideal power control with the round robin scheduler. Since the total rate of the ten users is $144 \times 5 + 12.2 \times 5 = 781$ kb/s, we set the processing gain $\gamma_{proc} = floor(3840/781) = 4$. We generate the channel power gain $g_k(t)$ for each user k and obtain the ideal power control by

$$P_0^{(k)}(t) = \frac{SINR_{target} \times (\sigma_n^2 + P_I(t))}{g_k(t) \times \gamma_{proc}}. \tag{204}$$

Denote $P_{avg}^{(k)}$ the time average of $P_0^{(k)}(t)$. For this simulation, the channel usage of ideal power control with the round robin scheduler is simply $781/(3840/\gamma_{proc}) = 0.81$.

In the second simulation, we simulate the K&H/RR scheduler and each user k has its transmit power fixed at $P_{avg}^{(k)}$ for fair comparison. Since the voice users and the video users have quite different QoS requirements, we apply the K&H/RR scheduler to the two groups separately like in Section 5.3.2. Assume $\{\zeta_{voice}, \beta_{voice}\}$ and $\{\zeta_{video}, \beta_{video}\}$ are allocated to the voice users and the video users, respectively. For simplicity, we employ either the K&H or the RR scheduler, depending on which one gives a smaller channel usage.

We use the sample path of the channel power gain $g_k(t)$ generated in the first simulation, also for fair comparison. At each frame t, we determine the processing gain $\gamma_{proc}(t)$ by

$$\gamma_{proc}^{(k)}(t) = ceiling\left(\frac{SINR_{target} \times (\sigma_n^2 + P_I(t))}{g_k(t) \times P_{avg}^{(k)}}\right). \tag{205}$$

where $ceiling(x)$ is the least integer that is not less than x. Then the transmission rate for a voice user k at frame t is given by $\varpi \times 3.84 \times 10^6/\gamma_{proc}^{(k)}(t)$, where $\varpi = \zeta_{voice}$ or β_{voice}, depending on the type of the scheduler. This rate adaptation also applies to the video users. From the simulation, the minimum channel usage is achieved when the voice users are scheduled by the RR

206

while the video users are scheduled by the K&H; the channel usages are 0.03 and 0.2, respectively. Hence, the total channel usage is 0.23, which is much smaller than that obtained in the case with ideal power control and the RR. This shows that substantial gain can be achieved by utilizing time diversity (the fixed power scheme) and multiuser diversity (the K&H), as opposed to the ideal power control and the RR, used in 3G systems.

7.5   Summary

In this chapter, we addressed some of the practical aspects of the effective capacity approach, namely, the effect of modulation and channel coding, and robustness against non-stationary channel gain processes.

We showed how to quantify the effect of practical modulation and channel coding on the effective capacity approach to QoS provisioning. We identified the important trade-off between power control and time diversity in QoS provisioning over fading channels, and proposed a novel time-diversity dependent power control scheme to leverage time diversity. Compared to ideal power control, an approximation of which is applied to the 3G wireless systems, our time-diversity dependent power control can achieve a factor of 2 to 4.7 capacity gain for mobile speeds between 0.11 m/s and 56 m/s.

Since time-varying mobile speeds, large scale path loss, shadowing, and very low mobility can cause severe QoS violations, it is essential to design QoS provisioning mechanisms to mitigate these channel conditions. Equipped with the newly developed time-diversity dependent power control and the effective capacity approach, we designed a power control and scheduling mechanism, which is robust in QoS provisioning against large scale fading and non-stationary small scale fading. With this mechanism, we proposed QoS provisioning algorithms for downlink and uplink transmissions, respectively. We also considered QoS provisioning for multiple users. Simulation results demonstrated the effectiveness and efficiency of our proposed algorithms in providing QoS guarantees under various channel conditions.

CHAPTER 8

ADAPTIVE QOS CONTROL FOR
WIRELESS VIDEO COMMUNICATION

8.1 Introduction

With technological advances and the emergence of personal communication era, the demand for wireless video communication is increasing rapidly. Compared with wired links, wireless channels are typically much more noisy and have both small-scale and large-scale fades [119], making the bit error rate very high. The resulting bit errors could have devastating effects on the video presentation quality. Thus, robust transmission of real-time video over wireless channels is a critical problem.

To address the above problem, we introduce adaptive QoS control for video communication over wireless channels. The objective of our adaptive QoS control is to achieve good perceptual quality and utilize network resources efficiently through adaptation to time-varying wireless channels. Our adaptive QoS control consists of optimal mode selection and delay-constrained hybrid ARQ. Optimal mode selection provides QoS support (*i.e.*, error resilience) on the compression layer while delay-constrained hybrid ARQ provides QoS support (*i.e.*, bounded delay and reliability) on the link layer. The main contributions of this chapter are: (1) an optimal mode selection algorithm which provides the best trade-off between compression efficiency and error resilience in rate-distortion (R-D) sense, and (2) delay-constrained hybrid ARQ which is capable of providing reliability for the compression layer while guaranteeing delay bound and achieving high throughput.

To facilitate the discussion of our adaptive QoS control, we first present an overall transport architecture for real-time video transmission over the wireless channel, which includes source rate adaptation, packetization, adaptive QoS control, and modulation. Since a wireless link is typically bandwidth-limited, it is essential for the source to adjust its transmission rate to the available bandwidth in the wireless channel. Therefore, source rate adaptation must be in place. Since bit-oriented syntax of a compressed video stream has to be converted into packets for transport over wireless channel, packetizing the compressed-video bit-stream is required. Adaptive QoS control is our primary weapon to combat bit errors under wireless environments. Finally, since modulation is required by any wireless communication, modulation is also an indispensable component of our architecture.

Under the above transport architecture, we discuss our adaptive QoS control. The first component is optimal mode selection. As we know, the effect of bit errors on the perceptual quality depends on the compression scheme used at the source, the wireless channel conditions and the error concealment scheme used at the receiver. High-compression coding algorithms usually employ inter-coding (*i.e.*, prediction) and variable length coding to achieve efficiency. Variable length coding is very susceptible to bit errors since a single bit error can corrupt a whole segment between two resynchronization markers in the compressed bit-stream [64], making the whole segment undecodable and resulting in discard of the video segment. For inter-coded video, discard of a video segment due to bit errors may degrade video quality over a large number of frames, until the next intra-coded frame is received. Intra-coding effectively stops error propagation, but at the cost of losing efficiency; inter-coding achieves compression efficiency, but at the risk of error propagation. Therefore, an optimal mode selection algorithm should be in place to achieve the best trade-off between efficiency and robustness.

The problem of mode selection has been well studied under the R-D framework [98, 123]. The previous approach to mode selection is to choose a mode that minimizes the quantization distortion between the original frame/macroblock and the reconstructed one under a given bit budget [98, 123], which is the so-called R-D optimized mode selection. We refer such R-D optimized mode selection as the classical approach. The classical approach is not able to achieve global optimality under the error-prone environment since it does not consider the channel characteristics and the receiver behavior. To remedy the drawback of the classical approach, we propose an end-to-end approach to R-D optimized mode selection. Different from the classical approach, our approach also considers the channel characteristics and the receiver behavior (*i.e.*, error concealment), in addition to quantization. Such an end-to-end approach was first introduced in [149] for Internet video communication. In this chapter, we show that such a methodology is also applicable to wireless environments with some modifications.

Previous work on optimal mode selection that considered wireless channel characteristics and error concealment has been reported in Ref. [32]. However, the distortion metrics introduced there are not accurate since the derivation of the distortion metrics was done at the block level. In addition, the wireless channel characteristics is dynamically changing, that is, the BER is varying from time to time. The scheme in Ref. [32] assumes that the BER is fixed and known *a priori*, which may not reflect the error behavior in the wireless channel. Thus, the scheme in Ref. [32] may not achieve optimality for dynamic wireless environments. This chapter addresses these problems by deriving the distortion metrics at the pixel level and employing feedback mechanism to deal with the time-varying nature of the wireless channel. We will show how to formulate the globally R-D optimized mode selection and how to derive accurate global distortion metrics under our architecture. Along the same direction, Zhang et al. [166] also derived distortion metrics at the pixel level. Our distortion metrics are similar to theirs but are different.

Besides the QoS support on the compression layer provided by optimal mode selection, QoS support for video communication can be provided on the link layer as well.

The second component of our adaptive QoS control is delay-constrained hybrid ARQ, which is an error-control mechanism on the link layer. There are two kinds of error-control mechanisms on the link layer, namely, FEC and ARQ. The principle of FEC is to add redundant information so that original message can be recovered in the presence of bit errors. The use of FEC is primarily because throughput can be kept constant and delay can be bounded under FEC. However, the redundancy ratio (the ratio of redundant bit number to total bit number) should be made large enough to guarantee target QoS requirements under the worst channel conditions. In addition, FEC is not adaptive to varying wireless channel condition and it works best only when the BER is stable. If the number of bit errors exceeds the FEC code's recovery capability, the FEC code cannot recover any portion of the original data. In other words, FEC is useless when the short-term BER exceeds the recovery capability of the FEC code. On the other hand, when the wireless channel is in good state (*i.e.*, the BER is very small), using FEC will cause unnecessary overhead and waste bandwidth. Different from FEC, ARQ is adaptive to varying wireless channel condition. When the channel is in good state, no retransmissions are required and no bandwidth is wasted. Only when the channel condition becomes poor, the retransmissions will be used to recover the errors. However, adaptiveness and efficiency of ARQ come with the cost of unbounded delay. That is, in the worst case, a packet may be retransmitted in unlimited times to recover bit errors.

To deal with the problems associated with FEC and ARQ, truncated type-II hybrid ARQ schemes have been proposed [82, 164]. Different from conventional type-II hybrid ARQ [55, 70, 81, 140], the truncated type-II hybrid ARQ has the restriction of maximum number of transmissions for a packet. Due to the maximum number of transmissions, delay can be bounded. The truncated type-II hybrid ARQ combines the good features of FEC and ARQ: bounded delay and adaptiveness. However, the maximum number of transmissions N_m

is assumed to be fixed and known *a priori* [82, 164], which may not reflect the time-varying nature of delay. If N_m is set too large, retransmitted packets may arrive too late for play-out and thereby be discarded, resulting in waste of bandwidth; if N_m is set too small, the perceptual quality will be reduced due to unrecoverable errors that could have been corrected with more retransmissions. We address this problem by introducing delay-constrained hybrid ARQ. In our delay-constrained hybrid ARQ, the receiver makes retransmission requests in an intelligent way: when errors in the received packet are detected, the receiver decides whether to send a retransmission request according to the delay bound of the packet. Our delay-constrained hybrid ARQ is shown to be capable of achieving bounded delay, adaptiveness, and efficiency.

Our adaptive QoS control is aimed at providing good visual quality under time-varying wireless channels. By combining the best features of error-resilient source encoding, FEC and delay-constrained retransmission, the proposed adaptive QoS control is shown to be capable of achieving better quality for real-time video under varying wireless channel conditions while utilizing network resources efficiently.

The remainder of this chapter is organized as follows. Section 8.2 sketches the overall transport architecture for real-time video communication over the wireless channel. In Section 8.3, we describe our optimal mode selection algorithm. Section 8.4 presents the delay-constrained hybrid ARQ. In Section 8.5, we use simulation results to demonstrate the effectiveness of our adaptive QoS control. Section 8.6 concludes this chapter.

8.2 An Architecture for Real-time Video Communication over a Wireless Channel

In this section, we describe an architecture for real-time video communication over a wireless channel. We organize this section as follows. In Section 8.2.1, we overview our architecture. From Sections 8.2.2 through 8.2.4, we briefly describe key components in our architecture.

8.2.1 Overview

Figure 77 shows our architecture for one-to-one video communication over the wireless channel. On the sender side, raw bit-stream of live video is encoded by a rate-adaptive video encoder, which also employs optimal mode selection. After this stage, the compressed video bit-stream is first packetized and then passed to CRC & RCPC encoder. Cyclic redundancy check (CRC) code and rate-compatible punctured convolutional (RCPC) codes are used as the error detection and correction codes. After modulation, packets are transmitted over a wireless channel. At the receiver, the video sequence is reconstructed in the reverse manner shown in Fig. 77.

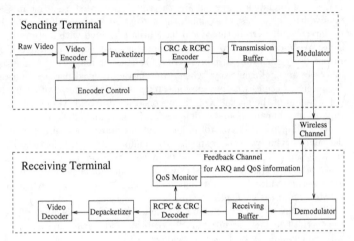

Figure 77: An architecture for MPEG-4 video over the wireless channel.

Under our architecture, a QoS monitor is kept at the receiver side to infer channel status based on the behavior of the arriving packets, *e.g.*, bit errors on the link layer, delay, and macroblock error ratio (MER). Link-layer bit errors and delay are used by hybrid ARQ to decide whether a retransmission should be requested. In addition, MER is periodically sent back to the source; the video encoder makes the optimal mode selection based on the returned MER. The feedback channel conveys two kinds of messages to the source: retransmission request and MER.

We use rate-adaptive video encoder [150] so that the rate of compressed video (denoted by $R_s(t)$) can match the available bandwidth of the wireless link (denoted by $R_a(t)$), which is time-varying due to probabilistic bit errors and retransmissions. In [150], we described how to adapt $R_s(t)$ to the target $R_a(t)$. Here, $R_a(t)$ is meant to be the maximum data rate that a video source can transmit at time t without error. It does not include the overheads due to ARQ and channel codes.

Next, we briefly describe three key components in our system: optimal mode selection, resynchronization markers, and delay-constrained hybrid ARQ.

8.2.2 Optimal Mode Selection

The problem of classical R-D optimized mode selection is to find the mode that minimizes the quantization distortion D_q for a given macroblock (MB), subject to a constraint. The classical R-D optimized mode selection is optimal with respect to quantization distortion. However, under wireless environments, where packets may get discarded due to unrecoverable errors, the classical R-D optimized mode selection is not optimal with respect to the distortion D_r, which measures the difference between the original MB at the source and the reconstructed one at the receiver. This is because the classical R-D optimized mode selection does not consider the channel characteristics and receiver behavior, which also affect the distortion D_r.

In Section 8.3, we will present an optimal mode selection algorithm for MPEG-4 video over wireless channels. Our algorithm is capable of achieving best trade-off between error resilience and efficiency in an R-D sense.

8.2.3 Resynchronization Markers

Resynchronization markers are used to prevent error propagation in the bit-stream when errors occur [135]. Resynchronization markers are specially designed bit patterns that are usually placed at approximately regular intervals in the video bit stream. The function of these markers is to divide the compressed video bit stream into segments that are as independent of each other as possible. By searching for these markers, the decoder can reliably locate each segment without actually decoding the packet, and thereby prevent error propagation across different segments separated by markers. Each data segment of the bit stream should generally contain one or several complete logical entities of video information (*e.g.*, MB's) so that the decrease in coding efficiency (due to lack of exploiting dependencies between segments) can be minimized. The length of each segment is usually chosen to achieve a good trade-off between the overhead introduced by the markers, and reliability of the detection of markers when errors occur.

Our resynchronization scheme takes the video packet approach adopted by MPEG-4 [135]. Under the video packet approach, resynchronization markers are inserted to the bit-stream periodically. Specifically, if the number of bits after the current resynchronization marker exceeds a predetermined threshold, a new resynchronization marker is inserted at the start of the next MB.

After insertion of resynchronization markers, the bit-stream is passed to packetizer. We use a fixed packet size for packetization. After packetization, the video packet is passed to the link layer (CRC & RCPC encoder). For each video packet, the CRC & RCPC encoder generates a link-layer packet by using CRC/RCPC codes.

213

8.2.4 Delay-constrained Hybrid ARQ

To provide adaptive QoS control on the link layer, delay-constrained hybrid ARQ is performed in our architecture. Like other hybrid ARQ schemes [82, 164], our hybrid ARQ is also a combination of FEC and ARQ. In our hybrid ARQ, CRC and RCPC are employed as the FEC codes and RCPC offers variable rates to adapt to the time-varying wireless channel.

Our hybrid ARQ works as follows. When errors in the received link-layer packet is detected, the receiver decides whether to send a retransmission request according to the delay bound of the packet. If the delay requirement cannot be met, the request will not be sent. That is, the retransmission is truncated, instead of repeating retransmissions until a successful reception occurs. If the delay requirement can be met, a retransmission request will be sent to the source. Upon a retransmission request, the source only transmits the necessary incremental redundancy bits.

In Section 8.4, we will present our delay-constrained hybrid ARQ that is capable of providing reliability while achieving high throughput.

Table 11 lists the notations used in this chapter. Under QCIF format, N_f is 99 and N_h is 98.

8.3 Optimal Mode Selection

As discussed in Section 8.2.2, the classical approach to mode selection is not able to achieve global optimality under the error-prone environment since it does not consider the channel conditions and the receiver behavior. In this section, we present an optimal mode selection algorithm.

We organize this section as follows. Section 8.3.1 discusses global distortion from an end-to-end perspective and set a stage for derivation of the global distortion metrics. In Sections 8.3.2, we derive the global distortion metrics. In Section 8.3.3, we design an algorithm for optimal mode selection based on the global distortion metrics.

8.3.1 An End-to-End Perspective

An end-to-end approach to R-D optimized mode selection was introduced in Ref. [149] for Internet video communication. With the end-to-end approach, the distortion D_r is assumed to be the difference between the original MB at the source and the reconstructed one at the receiver. Under wireless environments, the distortion D_r is a random variable, which may take the value of either (1) the quantization distortion D_q plus the distortion D_{ep} caused by error propagation, or (2) distortion D_c caused by errors due to error concealment. We define the *global* distortion D as the expectation of the random variable

Table 11: Notations.

N_f	:	the total number of MBs in a frame
N_h	:	the highest location number of MBs in a frame ($N_h = N_f - 1$)
N_G	:	the number of MBs in a group of blocks (GOB)
F_i^n	:	the MB at location i in frame n
$\bar{F}_i^n$	:	the coded MB at location i in frame n.
$F_{\bar{i}}^n$	:	the MB (at location $\bar{i}$ in frame n) which is above F_i^n, if it exists.
$F_{\tilde{i}}^n$	:	the MB (at location $\tilde{i}$ in frame n) which is below F_i^n, if it exists.
$\mathcal{G}^n$	:	the set of macroblocks F_i^n ($i \in [0, N_h]$) that does not have $F_{\bar{i}}^n$ or does not have $F_{\tilde{i}}^n$.†
f_{ij}^n	:	the original value of pixel j in F_i^n (raw data).
$\hat{f}_{ij}^n$	:	the value of reconstructed pixel j in F_i^n at the encoder.
$\tilde{f}_{ij}^n$	:	the value of reconstructed pixel j in F_i^n at the receiver.
e_{ij}^n	:	the prediction error of pixel j in inter-coded F_i^n.
$\tilde{e}_{ij}^n$	:	the reconstructed prediction error of pixel j in inter-coded F_i^n.
$\hat{f}_{uv}^{n-1}$	:	the value of reconstructed pixel v in F_u^{n-1} for prediction of f_{ij}^n.
$\tilde{f}_{ml}^{n-1}$	:	the value of reconstructed pixel l in F_m^{n-1} to replace $\hat{f}_{ij}^n$ due to the error concealment.
$\mathcal{I}$	:	the set of coding modes (*i.e.*, $\mathcal{I} = \{\text{intra, inter}\}$).
M_i^n	:	the mode selected to code macroblock F_i^n ($M_i^n \in \mathcal{I}$).
$\bar{\lambda}$	:	the Lagrange multiplier.

† For example, in QCIF, an MB in the first GOB does not have an MB above it; an MB in the last GOB does not have an MB below it.

D_r. That is,

$$D = E\{D_r\}, \tag{206}$$

where D_r takes the value of $(D_q + D_{ep})$ or D_c with certain probability, which is determined by channel error characteristics.

Then, the problem of globally R-D optimized mode selection is to find the mode that minimizes the global distortion D for a given MB, subject to a constraint R_c on the number of bits used. For the MB at location i in frame n (denoted by F_i^n), this constrained problem reads as follows:

$$\min_{M_i^n} D(M_i^n) \quad \text{subject to} \quad R(M_i^n) \le R_c, \tag{207}$$

where $D(M_i^n)$ and $R(M_i^n)$ denote the *global* distortion and bit budget for macroblock F_i^n with a particular mode M_i^n, respectively.

In terms of mean squared error (MSE), we define the global distortion metric for macroblock F_i^n as follows:

$$\text{MSE}(F_i^n) = \frac{E\left\{\sum_{j=1}^{256}(f_{ij}^n - \hat{f}_{ij}^n)^2\right\}}{256} \tag{208}$$

where f_{ij}^n is the original value of pixel j in F_i^n (raw data), and $\hat{f}_{ij}^n$ is the value of reconstructed pixel j in F_i^n at the receiver.

The global distortion is affected by three factors: source behavior, channel characteristics, and receiver behavior, which are described in Sections 8.3.1.1 to 8.3.1.3, respectively.

8.3.1.1 Source Behavior

The source behavior, *i.e.*, quantization, have impact on global distortion. This is because quantization error can cause image distortion.

A block diagram of the video encoder is depicted in Fig. 78. The switches represent the two different paths for the intra- and inter-mode.

Under the intra mode, the raw video f_{ij}^n is transformed by discrete cosine transform (DCT), quantized and coded by run length coding (RLC). The resulting information, as well as coding parameters such as the coding mode and the quantization parameter (QP), is coded by variable length coding (VLC). Then the compressed video stream is formed by the multiplexer. At the same time, the pixel is reconstructed at the encoder for the prediction used by the next frame. The value of the reconstructed pixel is $\tilde{f}_{ij}^n$.

Figure 78 also illustrates the case when the encoder is operating under the inter-mode. Under such mode, the raw video f_{ij}^n is first predicted from

216

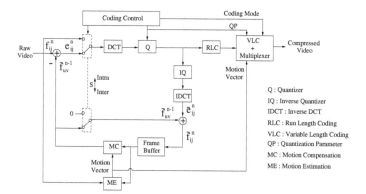

Figure 78: Block diagram of the video encoder working under the inter mode.

the motion-compensated pixel in the previous frame, $\tilde{f}_{uv}^{n-1}$, resulting in the prediction error, e_{ij}^n. The prediction error, e_{ij}^n, is DCT-transformed, quantized and coded by RLC. The resulting information, as well as coding parameters such as the coding mode, the motion vector and the QP, is coded by VLC. Then the compressed video stream is formed by the multiplexer. At the same time, the pixel is reconstructed at the encoder for the prediction used by the next frame. There are two steps for the reconstruction at the encoder. First, the prediction error is reconstructed, resulting in $\tilde{e}_{ij}^n$. Second, $\tilde{e}_{ij}^n$ is added to the predicted value $\tilde{f}_{uv}^{n-1}$, resulting in the reconstructed value, $\tilde{f}_{ij}^n$. That is, the pixel is reconstructed by $\tilde{f}_{ij}^n = \tilde{e}_{ij}^n + \tilde{f}_{uv}^{n-1}$.

8.3.1.2 Wireless Channel Characteristics

Wireless channel characteristics (*i.e.*, error characteristics) has great impact on the global distortion. Since only MB error ratio is relevant in deriving our global distortion metrics, we are primarily concerned with compression-layer error characteristics rather than link-layer error characteristics.

Under our architecture, delay-constrained hybrid ARQ is used to recover link-layer bit errors. If the bit errors in a link-layer packet cannot be completely recovered before its delay bound, the corrupted packet will be discarded. Then the MPEG-4 decoder uses error concealment to conceal the MB's associated with the corrupted link-layer packet. We will address error concealment in Section 8.3.1.3.

There are two ways to obtain MER, namely, analysis and measurement.

217

We can use analysis like Ref. [164] to derive MER. But the analysis is very complicated and we can only get an approximation of MER (*e.g.*, an upper bound). The approximation is typically only applicable to a certain channel model and a certain modulation method. For example, the upper bound on packet error ratio in Ref. [164] is only applicable to Rayleigh fading channel model and $\pi/4$-DQPSK modulation with coherent detection.

Due to the complexity and inflexibility of the analytical method, we choose a measurement approach in our system. That is, we measure the numbers of erroneous/discarded MB's and correct MB's at the receiver. That is, MER, denoted by p_d, is given by

$$MER = p_d = \frac{N_d}{N_d + N_c},\tag{209}$$

where N_d is the number of erroneous/discarded MB's and N_c the number of correct MB's.

8.3.1.3 Receiver Behavior

Receiver behavior, *i.e.*, error concealment, also affects the global distortion. In our system, we only consider such an error concealment scheme: the corrupted MB is replaced with the MB from the previous frame pointed by a motion vector. The motion vector of the corrupted MB is copied from one of its neighboring MB (which is above or below the corrupted MB) when available, otherwise the motion vector is set to zero. Specifically, if the MB above the corrupted MB is received correctly, the motion vector of the corrupted MB is copied from the MB above it; if the MB above the corrupted MB is also damaged and the MB below the corrupted MB is received correctly, the motion vector of the corrupted MB is copied from the MB below it; otherwise the motion vector is set to zero.

Figure 79 shows a block diagram of the video decoder, in which three switches represent different scenarios as follows.

- *Switch SW1:* represents the two different paths for the intra- and inter-mode.

- *Switch SW2:* represents the two different paths for the two cases as follows.

  − the case where the coded MB at location i in frame n (denoted by $\bar{F}_i^n$) is received correctly.

  − the case where $\bar{F}_i^n$ is corrupted.

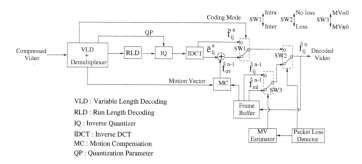

Figure 79: Block diagram of the video decoder.

- *Switch SW3:* represents the two different paths for the two cases as follows.

  - the case where the estimated motion vector of the corrupted F_i^n is set to zero.
  - the case where the estimated motion vector of the corrupted F_i^n is copied from its neighboring macroblock $F_{\bar{i}}^n$ or $F_{\tilde{i}}^n$, where $F_{\bar{i}}^n$ denotes the MB (at location $\bar{i}$ in frame n) which is above F_i^n and $F_{\tilde{i}}^n$ denotes the MB (at location $\tilde{i}$ in frame n) which is below F_i^n.

The compressed video stream is first demultiplexed by the demultiplexer and decoded by variable length decoding (VLD). The resulting coding mode information will control switch SW1. The resulting QP will control the inverse quantizer (IQ). The resulting motion vector is used for motion compensation. The information containing DCT coefficient will be decoded by run length decoding (RLD), then inversely quantized, and inversely DCT-transformed.

There are three cases for the reconstructed pixel at the receiver as follows.

- *Case (i):* $\bar{F}_i^n$ is received correctly. If F_i^n is intra-coded, then we have $\hat{f}_{ij}^n = \tilde{f}_{ij}^n$, which is illustrated in Fig. 79 with state {SW1: Intra, SW2: No loss, SW3: don't care}. If F_i^n is inter-coded, then we have

$$\hat{f}_{ij}^n = \tilde{e}_{ij}^n + \hat{f}_{uv}^{n-1}, \tag{210}$$

  which is illustrated in Fig. 79 with state {SW1: Inter, SW2: No loss, SW3: don't care}.

219

- *Case (ii):* $\bar{F}_i^n$ is corrupted and the neighboring MB ($\bar{F}_{\bar{i}}^n$ or $\bar{F}_{\tilde{i}}^n$) has been received correctly. Then we have $\hat{f}_{ij}^n = \hat{f}_{ml}^{n-1}$, which is illustrated in Fig. 79 with state {SW1: don't care, SW2: Loss, SW3: MV$\neq$0}.

- *Case (iii):* $\bar{F}_i^n$, $\bar{F}_{\bar{i}}^n$, and $\bar{F}_{\tilde{i}}^n$ are all corrupted. Then we have $\hat{f}_{ij}^n = \hat{f}_{ij}^{n-1}$, which is illustrated in Fig. 79 with state {SW1: don't care, SW2: Loss, SW3: MV=0}.

So far we have examined three key components that affect the global distortion. In the next section, we derive the global distortion metrics based on the three factors discussed in this section.

8.3.2 Global Distortion Metrics

This section derives the global distortion metrics, which will be used by our optimal mode selection in Section 8.3.3.

Without loss of generality, we consider the distortion for macroblock s in frame N, where s ($s \in [0, N_h]$) is the location number and N ($N \geq 0$) is the frame number. Note that the sequence number for both frame and packet start from zero. Denote $\mathcal{G}^n$ the set of macroblocks F_i^n that does not have $F_{\bar{i}}^n$ or does not have $F_{\tilde{i}}^n$. For example, in QCIF, an MB in the first GOB does not have an MB above it; an MB in the last GOB does not have an MB below it.

The following proposition shows how to compute MSE for the intra-coded MB under our architecture.

Proposition 8.1 *Assume the packet loss at the compression layer follows Bernoulli distribution with probability p_d. Under our architecture, the MSE for the intra-coded MB at location s of frame N ($N > 0$) is given by*

$$MSE(F_s^N, intra) = \frac{\sum_{j=1}^{256}\left(\left(f_{sj}^N\right)^2 - 2 \times f_{sj}^N \times E\{\hat{f}_{sj}^N\} + E\left\{\left(\hat{f}_{sj}^N\right)^2\right\}\right)}{256},$$

(211)

where

$$E\{\hat{f}_{ij}^n\} = \begin{cases} (1-p_d) \times \tilde{f}_{ij}^n + p_d \times (1-p_d) \times E\{\hat{f}_{ml}^{n-1}\} + p_d^2 \times E\{\hat{f}_{ij}^{n-1}\} & \text{if } F_i^n \in \mathcal{G}^n \\ \\ (1-p_d) \times \tilde{f}_{ij}^n + p_d \times (1-p_d^2) \times E\{\hat{f}_{ml}^{n-1}\} + p_d^3 \times E\{\hat{f}_{ij}^{n-1}\} & \text{if } F_i^n \notin \mathcal{G}^n \end{cases}$$

(212)

$$
E\left\{\left(\hat{f}_{ij}^n\right)^2\right\} =
\begin{cases}
(1-p_d) \times \left(\tilde{f}_{ij}^n\right)^2 + p_d \times (1-p_d) \times E\left\{\left(\hat{f}_{ml}^{n-1}\right)^2\right\} + \\[2mm]
p_d^2 \times E\left\{\left(\hat{f}_{ij}^{n-1}\right)^2\right\} & \text{if } F_i^n \in \mathcal{G}^n \\[6mm]
(1-p_d) \times \left(\tilde{f}_{ij}^n\right)^2 + p_d \times (1-p_d^2) \times E\left\{\left(\hat{f}_{ml}^{n-1}\right)^2\right\} + \\[2mm]
p_d^3 \times E\left\{\left(\hat{f}_{ij}^{n-1}\right)^2\right\} & \text{if } F_i^n \notin \mathcal{G}^n
\end{cases}
, \tag{213}
$$

and $E\{\hat{f}_{ij}^0\} = \tilde{f}_{ij}^0$ and $E\left\{\left(\hat{f}_{ij}^0\right)^2\right\} = \left(\tilde{f}_{ij}^0\right)^2$.

For a proof of Proposition 8.1, see Appendix A.18.

The following proposition shows how to compute MSE for the inter-coded MB under our architecture.

Proposition 8.2 *Assume the packet loss at the compression layer follows Bernoulli distribution with probability p_d. Under our architecture, the MSE for the inter-coded MB at location s of frame N ($N > 0$) is given by*

$$
MSE(F_s^N, inter) = \frac{\sum_{j=1}^{256}\left(\left(f_{sj}^N\right)^2 - 2 \times f_{sj}^N \times E\{\hat{f}_{sj}^N\} + E\left\{\left(\hat{f}_{sj}^N\right)^2\right\}\right)}{256}, \tag{214}
$$

where

$$
E\{\hat{f}_{ij}^n\} =
\begin{cases}
(1-p_d) \times (\tilde{e}_{ij}^n + E\{\hat{f}_{uv}^{n-1}\}) + p_d \times (1-p_d) \times E\{\hat{f}_{ml}^{n-1}\} + \\[2mm]
p_d^2 \times E\{\hat{f}_{ij}^{n-1}\} & \text{if } F_i^n \in \mathcal{G}^n \\[6mm]
(1-p_d) \times (\tilde{e}_{ij}^n + E\{\hat{f}_{uv}^{n-1}\}) + p_d \times (1-p_d^2) \times E\{\hat{f}_{ml}^{n-1}\} + \\[2mm]
p_d^3 \times E\{\hat{f}_{ij}^{n-1}\} & \text{if } F_i^n \notin \mathcal{G}^n
\end{cases}
, \tag{215}
$$

$$
E\left\{\left(\hat{f}_{ij}^{n}\right)^{2}\right\} =
\begin{cases}
(1-p_d) \times \left(\left(\bar{e}_{ij}^{n}\right)^{2} + 2 \times \bar{e}_{ij}^{n} \times E\{\hat{f}_{uv}^{n-1}\} + E\left\{\left(\hat{f}_{uv}^{n-1}\right)^{2}\right\}\right) + \\[2mm]
p_d \times (1-p_d) \times E\left\{\left(\hat{f}_{ml}^{n-1}\right)^{2}\right\} + p_d^{2} \times E\left\{\left(\hat{f}_{ij}^{n-1}\right)^{2}\right\} \qquad \text{if } F_i^{n} \in \mathcal{G}^{*} \\[4mm]
(1-p_d) \times \left(\left(\bar{e}_{ij}^{n}\right)^{2} + 2 \times \bar{e}_{ij}^{n} \times E\{\hat{f}_{uv}^{n-1}\} + E\left\{\left(\hat{f}_{uv}^{n-1}\right)^{2}\right\}\right) + \\[2mm]
p_d \times (1-p_d^{2}) \times E\left\{\left(\hat{f}_{ml}^{n-1}\right)^{2}\right\} + p_d^{3} \times E\left\{\left(\hat{f}_{ij}^{n-1}\right)^{2}\right\} \qquad \text{if } F_i^{n} \notin \mathcal{G}^{*}
\end{cases}
\tag{216}
$$

and $E\{\hat{f}_{ij}^{0}\} = \tilde{f}_{ij}^{0}$ and $E\left\{\left(\hat{f}_{ij}^{0}\right)^{2}\right\} = \left(\tilde{f}_{ij}^{0}\right)^{2}$.

For a proof of Proposition 8.2, see Appendix A.19.

8.3.3 Optimal Mode Selection

Given the channel characteristics (obtained from measurement described in Section 8.3.1.2), we can design a globally R-D optimized mode selection algorithm for our architecture.

Consider a group of blocks (GOB) or slice denoted by $\mathcal{F}_g^{n} = (F_g^{n}, \cdots, F_{g+N_G-1}^{n})$, where N_G is the number of MBs in a GOB/slice. Assume each MB in $\mathcal{F}_g^{n}$ can be coded using only one of the two modes in set $\mathcal{I}$. Then for a given GOB/slice, the modes assigned to the MBs in $\mathcal{F}_g^{n}$ are given by the N_G-tuple, $\mathcal{M}_g^{n} = (M_g^{n}, \cdots, M_{g+N_G-1}^{n}) \in \mathcal{I}^{N_G}$. The problem of globally R-D optimized mode selection is to find the combination of modes that minimizes the distortion for a given GOB/slice, subject to a constraint R_c on the number of bits used. This constrained problem can be formulated as

$$
\min_{\mathcal{M}_g^{n}} D(\mathcal{F}_g^{n}, \mathcal{M}_g^{n}) \quad \text{subject to} \quad R(\mathcal{F}_g^{n}, \mathcal{M}_g^{n}) \leq R_c,
\tag{217}
$$

where $D(\mathcal{F}_g^{n}, \mathcal{M}_g^{n})$ and $R(\mathcal{F}_g^{n}, \mathcal{M}_g^{n})$ denote the total distortion and bit budget, respectively, for the GOB/slice $\mathcal{F}_g^{n}$ with a particular mode combination $\mathcal{M}_g^{n}$.

The constrained minimization problem in (217) can be converted to an unconstrained minimization problem by Lagrange multiplier technique as follows:

$$
\sum_{i=g}^{g+N_G-1} \min_{M_i^{n}} J(F_i^{n}, M_i^{n}) = \sum_{i=g}^{g+N_G-1} \min_{M_i^{n}} \{D(F_i^{n}, M_i^{n}) + \tilde{\lambda} R(F_i^{n}, M_i^{n})\},
\tag{218}
$$

where the global distortion $D(F_i^{n}, M_i^{n})$ can be expressed by the formulae we derived in Section 8.3.2, according to the coding mode.

222

The problem of (218) is a standard R-D optimization problem and can be solved by the approaches described in [77, 107, 117, 144]. Different from these approaches, we use a simpler method to obtain $\tilde{\lambda}$.

Since a large $\tilde{\lambda}$ in the optimization problem of (218) can reduce the bit-count of the coded frame, we employ this nature in choosing $\tilde{\lambda}$. To be specific, at the end of frame n, we adjust $\tilde{\lambda}$ for frame $n + 1$ (i.e., $\tilde{\lambda}_{n+1}$) as follows:

$$\tilde{\lambda}_{n+1} = \frac{2 \cdot B_n + (\tilde{\beta} - B_n)}{B_n + 2 \cdot (\tilde{\beta} - B_n)} \cdot \tilde{\lambda}_n \qquad (219)$$

where B_n is the current buffer occupancy at the end of frame n and $\tilde{\beta}$ is the buffer size. $\tilde{\lambda}_n$ is initialized by a preset value $\tilde{\lambda}_0$. The adjustment in Eq. (219) is to keep the buffer occupancy at the middle level to reduce the chance of buffer overflow or underflow. In other word, Eq. (219) also achieves the objective of rate control.

8.4 Hybrid ARQ

As mentioned in Section 8.1, FEC schemes offer a constant throughput but their reliability degrades as the channel condition becomes poor. ARQ achieves high efficiency and high throughput under good channel conditions, but the throughput is shrunken rapidly due to increase of retransmissions under poor channel conditions. The performance of an ARQ scheme can be improved by embedding FEC capability in it, which is so-called *hybrid ARQ*. The advantage of using FEC with ARQ is that many errors can be corrected by the FEC, and retransmissions are requested only when errors are unrecoverable by the FEC. This results in a reduction in the overall retransmission frequency and hence improves throughput during poor channel conditions.

There are two classes of hybrid ARQ schemes: type-I and type-II [81]. A general type-I hybrid ARQ scheme uses a block code (*e.g.*, a Reed-Solomon code), which is designed for simultaneous error correction and error detection. When a received packet is detected in error, the receiver first attempts to locate and correct the errors. If the number of bit errors is within the error-correcting capability of the code, the errors will be corrected and the decoded message will be passed to the user (*e.g.*, compression layer in our system). If an uncorrectable error pattern is detected, the receiver discards the received packet and requests a retransmission.

In the following sections, we describe a general type-II hybrid ARQ scheme and our delay-constrained hybrid ARQ in detail.

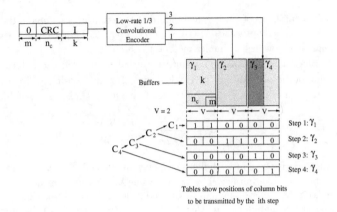

Figure 80: Structure of the RCPC codes.

8.4.1 General Type II Hybrid ARQ

A general type-II hybrid ARQ scheme uses a family of codes (*e.g.*, RCPC codes), each of which has different correction capability. Let $C_1 > C_2 > \cdots > C_M$ denote the M rates offered by a family of RCPC codes which are obtained from a low rate C_M (*e.g.*, 1/3) code, called the parent code. The parent code is an ordinary convolutional code. Figure 80 shows the structure of the RCPC codes with a 1/3 convolution code as a parent code. In Fig. 80, the puncture period V is 2 and the rates are 1, 1/2, 2/5, and 1/3. The corresponding puncturing matrices $\mathbf{P}$ are given by

$$\mathbf{P}_{1/1} = \begin{bmatrix} 1 & 1 \\ 0 & 0 \\ 0 & 0 \end{bmatrix}, \quad \mathbf{P}_{1/2} = \begin{bmatrix} 1 & 1 \\ 1 & 1 \\ 0 & 0 \end{bmatrix}, \quad \mathbf{P}_{2/5} = \begin{bmatrix} 1 & 1 \\ 1 & 1 \\ 1 & 0 \end{bmatrix}, \quad \mathbf{P}_{1/3} = \begin{bmatrix} 1 & 1 \\ 1 & 1 \\ 1 & 1 \end{bmatrix}.$$

(220)

As shown in Fig. 80, a number n_c of CRC bits for error detection are attached to the information packet (k bits). Tail bits m zeros are also added to properly terminate the encoder memory and decoder trellis. Each $(k+n_c+m)$-bit segment is encoded with the parent code C_M encoder.

There are three buffers corresponding to three output ports of the 1/3 convolutional encoder. Each buffer contains $(k + n_c + m)$ bits. Each row of the buffers has two bits since the puncturing period is two. The tables in

Fig. 80 are generated from the puncturing matrices $\mathbf{P}_{1/1}$ to $\mathbf{P}_{1/3}$ in Eq. (220). Specifically, the first row corresponds to $\mathbf{P}_{1/1}$; the first row and the second row correspond to $\mathbf{P}_{1/2}$; the first three rows correspond to $\mathbf{P}_{2/5}$; the four rows correspond to $\mathbf{P}_{1/3}$.

In Fig. 80, each row in the tables shows the positions of column bits to be transmitted by the i-th step. As a result, we get packets $\gamma_1, \gamma_2, \gamma_3$, and γ_4, which correspond to each row, respectively. In other words, packet γ_1 is generated by code C_1; packets γ_1 and γ_2 are generated by code C_2; packets γ_1, γ_2, and γ_3 are generated by code C_3; packets $\gamma_1, \gamma_2, \gamma_3$ and γ_4 are generated by code C_4.

After the encoding, the sender first sends packet γ_1 to the receiver. If packet γ_1 is received with unrecoverable errors, the receiver generates an ARQ request (note that packet γ_1 is not discarded, which is different from a type-I hybrid ARQ scheme). Then the sender transmits packet γ_2. Upon receipt of packet γ_2, the receiver decodes the combination of packets γ_1 and γ_2 by using code C_2. This procedure of retransmission is continued until decoding process results in no errors being detected. For a truncated type-II hybrid ARQ scheme, the procedure of retransmission is continued until decoding process results in no errors being detected or the maximum number of retransmissions is reached. For our delay-constrained hybrid ARQ, the procedure of retransmission is continued until decoding process results in no errors being detected or the delay bound is violated.

8.4.2 Delay-constrained Hybrid ARQ

Unlike data, video transmission can tolerate certain errors, but it has stringent delay requirements. This is because the packets arriving later than their playout time will not be displayed. Too many retransmissions could incur excessive delay, which exceeds the delay requirement of real-time video. For this reason, we introduce delay constraint to limit the number of retransmissions and satisfy the QOS requirements of video communications.

Our delay-constrained hybrid ARQ has the same transmission process as a general type-II hybrid ARQ scheme described in Section 8.4.1. The difference between them is the procedure of retransmission. Next, we describe the retransmission mechanisms for our delay-constrained hybrid ARQ.

Based on who determines whether to send and/or respond to a retransmission request, we design three delay-constrained retransmission mechanisms, namely, receiver-based, sender-based, and hybrid control.

Receiver-based control. The objective of the receiver-based control is to minimize the requests of retransmission that will not arrive timely for display. Under the receiver-based control, the receiver executes the following algorithm.

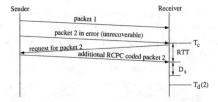

Figure 81: Timing diagram for receiver-based control.

When the receiver detects unrecoverable errors in a link-layer packet (corresponding
to information packet N):

> if $(T_c + RTT + D_s < T_d(N))$
>
>> send the request for retransmission of a repair packet (corresponding to
>> information packet N) to the sender;

where T_c is the current time, RTT is an estimated round trip time, D_s is a
slack term, and $T_d(N)$ is the time when packet N is scheduled for display.
The slack term D_s could include tolerance of error in estimating RTT, the
sender's response time to a request, and/or the receiver's processing delay
(*e.g.*, decoding). If $T_c + RTT + D_s < T_d(N)$ holds, it is expected that the
retransmitted packet will arrive timely for display. The timing diagram for
receiver-based control is shown in Fig. 18.

Since RTT may not change in wireless environment, RTT can be obtained
during set-up phase through a test. Specifically, the source sends out a packet
with a timestamp; the receiver sends back the packet when the packet is received; when the packet is received by the source, the source can determine
the RTT based on the difference between its current time and the timestamp
embedded in the packet.

Sender-based control. The objective of the sender-based control is to suppress retransmission of packets that will miss their display time at the receiver.
Under the sender-based control, the sender executes the following algorithm.

When the sender receives a request for retransmission of a repair packet (correspond
to information packet N):

> if $(T_c + T_o + D_s < T'_d(N))$
>
>> retransmit a repair packet to the receiver

226

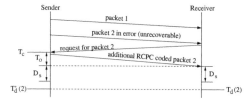

Figure 82: Timing diagram for sender-based control.

where T_o is the estimated one-way trip time (from the sender to the receiver), and $T'_d(N)$ is an estimate of $T_d(N)$. To obtain $T'_d(N)$, the receiver has to feedback $T_d(N)$ to the sender. Then, based on the differences between the sender's system time and the receiver's system time, the sender can derive $T'_d(N)$. The slack term D_s may include error terms in estimating T_o and $T'_d(N)$, as well as tolerance in the receiver's processing delay (*e.g.*, decoding). If $T_c + RTT + D_s < T'_d(N)$ holds, it can be expected that retransmitted packet will reach the receiver in time for display. The timing diagram for sender-based control is shown in Fig. 19.

Hybrid control. The objective of the hybrid control is to minimize the request of retransmissions that will not arrive for timely display, and to suppress retransmission of the packets that will miss their display time at the receiver. The hybrid control is a simple combination of the sender-based control and the receiver-based control. Specifically, the receiver makes decisions on whether to send retransmission requests while the sender makes decisions on whether to disregard requests for retransmission. The hybrid control may achieve better performance at the cost of higher complexity.

Our delay-constrained hybrid ARQ could use any of the three retransmission mechanisms, *i.e.*, receiver-based, sender-based, and hybrid control. In addition, to maximize throughput, we use selective-reject ARQ[1] rather than stop-and-wait ARQ or go-back-N ARQ [122]. Under selective-reject ARQ, the receiver responds with an acknowledgment containing the sequence number (*e.g.*, N) if the Nth packet is received correctly. On the other hand, if the Nth packet is corrupted (there is no way to decode it and know which packet it is), the sender retransmits the incremental redundant packet (*e.g.*, γ_2) when the counter times out; before the counter times out, if an acknowledgment for the $(N+1)$th packet rather than the Nth is received, the sender retransmits

---

[1]Selective-reject ARQ uses sliding-window flow control. Each packet has a sequence number indicating the position of the packet in the sliding window. Selective-reject ARQ is also called selective-repeat ARQ [81].

the incremental redundant packet for the Nth. To achieve highest through-put for selective-reject ARQ, the sliding-window size needs to be large enough (with respect to the ratio of RTT to T_{trans}, where T_{trans} is the transmission time of the packet) and the time-out counter for a packet needs to be set to $T_{trans} + RTT$ (suppose the transmission time of the acknowledgment packet and the processing time are negligible) [122].

Like other hybrid ARQ schemes [164], our hybrid ARQ combines the advantages of ARQ and FEC. Furthermore, our delay-constrained hybrid ARQ is more adaptive to varying wireless channel conditions while guaranteeing delay bound.

Next, we evaluate the performance of our adaptive QoS control mechanism through simulations.

8.5 Simulation Results

In this section, we implement the architecture described in Section 8.2 on our simulator and perform a simulation study of video conferencing with MPEG-4. The purpose of this section is to demonstrate (1) the performance improvement of our approach over the classical approach for optimal mode selection, and (2) the performance improvement of our delay-constrained hybrid ARQ over the truncated Type-II hybrid ARQ.

8.5.1 Simulation Settings

We implement the architecture shown in Fig. 77. At the source side, we use the standard raw video sequence "Miss America" in QCIF format for the video encoder. The encoder employs the rate control described in [150] so that the output rate of MPEG-4 encoder can adapt to the varying available bandwidth $R_a(t)$. The threshold for inserting resynchronization marker is set to 612 bits to match the link-layer payload size k. After compression, the video bit-stream is chopped into packets by the packetizer with a fixed size k of 612 bits. Then the packets are passed to CRC & RCPC encoder. We use 16-bit CRC code to detect bit errors. The 16-bit CRC error detection code will miss error patterns in packets with a probability of the order 10^{-5} of independent identically distributed (i.i.d.) errors. By using more powerful CRC codes, the probability of undetected error can be made negligible at the cost of reduced throughput.

RCPC codes are used as the error correction codes. Two sets of RCPC codes are used with the ARQ schemes. Both sets are generated from a rate $1/4$ code with constraint length $K = 5$ and puncturing period $V = 8$ (see Table B.189 in Ref. [48])). The rates of Set $\mathcal{A}$ are 1, 5/6, 5/8, 5/10, 5/12, 5/14, 5/16,

Table 12: Simulation parameters.

MPEG-4 encoder	Frame rate	10 frames/s
	I-VOP refreshment period	50 frames
	Mode	Rectangle
	$\tilde{\lambda}_0$	1
	End-to-end delay bound	100 ms
	Resynchronization marker length	17 bits
	Threshold for resynchronization marker	612 bits
Packetizer	Information packet size k	612 bits
CRC & RCPC encoder	Number of bits for packet sequence number	8 bits
	Number of CRC bits n_c	16 bits
	Number of tail bits m	4 bits
Hybrid ARQ	Sliding-window size	256 packets
	Packet processing delay	0 s
	Data transmission rate	64 kb/s
Wireless channel	RTT	10 ms or 30 ms
	channel model	AWGN or Rayleigh fading

$5/17, 5/18, 5/19$, and $1/4$. The rates of Set $\mathcal{B}$ are $1, 5/10, 5/15$, and $1/4$. At the receiver side, if an erroneous link-layer packet cannot be recovered by RCPC at its playout time, the link-layer packet and its retransmitted redundancy will be discarded. The sliding-window size for ARQ is set to 256 (using 8 bits), which is large enough (with respect to the ratio of RTT to the packet transmission time, which is approximately 4). We only use receiver-based control for our delay-constrained hybrid ARQ. This is because receiver-based control has the same performance as that of hybrid control when the RTT estimated by the receiver is accurate.

In our simulations, we assume that the feedback channel is error-free. Finite buffers are available at the transmitter and receiver and are large enough for storing transmitted and received packets for the link-layer ARQ protocol. For the simulations, we employ two kinds of wireless channels, namely, additive white Gaussian noise (AWGN) channel and Rayleigh fading channel. Binary phase shift keying (BPSK) with coherent detection is used as the modulation/demodulation scheme with a carrier frequency of 1.9 GHz.

Table 12 lists the parameters used in our simulation. Under such simulation settings, we consider three different encoders for MPEG-4 video as follows.

Encoder A: employs the classical approach for R-D optimized mode selection.

Encoder B: implements the globally R-D optimized mode selection described in Section 8.3. However, feedback of MER is not employed.

Encoder C: implements the globally R-D optimized mode selection described in Section 8.3. Feedback of MER to the source is employed.

Next, we show our simulation results under an AWGN channel and a Rayleigh fading channel in Sections 8.5.2 and 8.5.3, respectively. Each simulation is conducted for 100-second simulation time.

8.5.2   Performance over an AWGN Channel

In this section, we evaluate the performance of optimal mode selection and hybrid ARQ under the setting of an AWGN channel.

8.5.2.1   Performance of Optimal Mode Selection

In this simulation, we evaluate the performance of different optimal mode selection schemes, namely, Encoder A, B, and C. For Encoder B, we set MER=1% for optimal mode selection. The round-trip time is set to 10 ms. The delay-constrained hybrid ARQ uses the rates of Set $\mathcal{A}$ for RCPC codes and the delay bound is set to be 100 ms since the frame rate of the MPEG-4 stream is 10 frames/sec. Figure 83 shows the average peak signal-to-noise ratios (PSNR's) of Y component (denoted by PSNR-Y) under an AWGN channel with different channel signal-to-noise ratios (SNR's).[2]   We observe that Encoder C achieves the best performance, Encoder B has the second best performance, and Encoder A performs the worst. This demonstrates that our approach achieves better performance than the classical one, even if feedback mechanism is not employed. In addition, feedback-based scheme (*i.e.*, Encoder C) achieves better performance than non-feedback-based scheme (*i.e.*, Encoder B) since Encoder C has more accurate channel information than Encoder B. Moreover, Figure 83 indicates that the lower the channel SNR, the larger the performance gain achieved by our proposed scheme (*i.e.*, Encoder C). This feature enables our scheme to improve visual presentation quality substantially when a mobile device works under low channel SNR (*e.g.*, between 0 to 5 dB).

---

[2]Specifically, the channel SNR here is E_b/N_0, where E_b is the bit energy and N_0 is the single-sided power spectral density of noise.

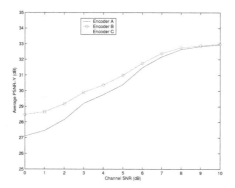

Figure 83: Performance of different video encoders.

8.5.2.2 Performance of Hybrid ARQ

To compare the performance of the truncated type-II hybrid ARQ [164] and our delay-constrained hybrid ARQ, we run our simulations under two scenarios where only Encoder C is employed.

In the first scenario, we set RTT=10 ms and the maximum number of retransmissions N_m=3 for the truncated hybrid ARQ. The truncated hybrid ARQ uses the rates of Set $\mathcal{B}$ for RCPC codes; our delay-constrained hybrid ARQ uses the rates of Set $\mathcal{A}$ for RCPC codes. Figure 84(a) shows that our delay-constrained hybrid ARQ achieves higher average PSNR-Y than the truncated hybrid ARQ, when the channel SNR is less than 7 dB. This is because when the channel SNR is less than 7 dB, an erroneous link-layer packet can be retransmitted 10 times under our delay-constrained hybrid ARQ without violating the delay bound (*i.e.*, 100 ms) while the truncated hybrid ARQ only allows 3 retransmissions, resulting in lower recovery probability. In other words, the truncated hybrid ARQ has higher probability of packet corruption. When the channel SNR is greater than or equal to 7 dB, at most 3 retransmissions are enough to recover all the bit errors. So, in this case, the two hybrid ARQ achieve the same average PSNR-Y.

In the second scenario, we set RTT=30 ms and the maximum number of retransmissions N_m=10 for the truncated hybrid ARQ. Both hybrid ARQ scheme use the rates of Set $\mathcal{A}$ for RCPC codes. Figure 84(b) shows that our delay-constrained hybrid ARQ achieves higher average PSNR-Y than the truncated hybrid ARQ, when the channel SNR is less than 7 dB. This is because when the channel SNR is less than 7 dB, an erroneous link-layer packet may be retransmitted up to 3 times under our delay-constrained hybrid ARQ

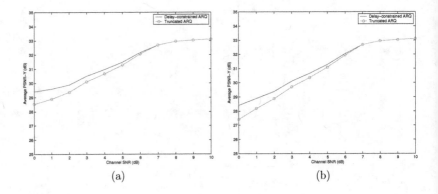

(a)   (b)

Figure 84: Average PSNR-Y for hybrid ARQ schemes: (a) the first scenario, (b) the second scenario.

without violating the delay bound (*i.e.*, 100 ms) while the truncated hybrid ARQ allows 10 retransmissions, resulting in waste of bandwidth due to discard of packets that missed their playout time. The more the bandwidth is wasted by the truncated hybrid ARQ, the less bandwidth MPEG-4 encoder can consume. This translates into lower average PSNR-Y for the truncated hybrid ARQ, as shown in Fig. 84(b). This is also confirmed by Fig. 85, which shows that the truncated hybrid ARQ achieves lower throughput than our delay-constrained hybrid ARQ, when the channel SNR is less than 7 dB. The throughput is defined as the ratio of the MPEG-4 encoding rate $R_s(t)$ to the link-layer transmission rate 64 kb/s. On the other hand, when the channel SNR is greater than or equal to 7 dB, at most 3 retransmissions are enough to recover all the bit errors. Hence, in this case, the two hybrid ARQ achieve the same throughput and the same average PSNR-Y.

The purpose of using the above two scenarios is to show that the truncated hybrid ARQ is not adaptive to varying *RTT*. In contrast, our delay-constrained hybrid ARQ is capable of adapting to varying *RTT* and achieving better performance than the truncated hybrid ARQ, for *real-time* video transmission over wireless channels.

8.5.3 Performance over a Rayleigh Fading Channel

In this section, we evaluate the performance of optimal mode selection and hybrid ARQ, under the setting of a Rayleigh fading channel. As compared to an AWGN channel, a Rayleigh fading channel with a specified Doppler spectrum

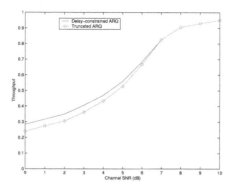

Figure 85: Throughput for hybrid ARQ schemes.

Table 13: Simulation parameters for a Rayleigh fading channel.

Sampling interval T_s	1/64000 second
Carrier frequency	1.9 GHz
Maximum Doppler rate f_m	5.3, 70, 211 Hz
	(corresponding to vehicle speed of 3, 40, 120 km/h,
	corresponding to coherence time T_c of 33.8, 2.6, 0.8 msec.)
Average SNR	0 to 20 dB

induces correlated error bursts, which are more difficult for the hybrid ARQ to recover.

As described in Section 4.4.1, Rayleigh flat-fading voltage-gains $h(t)$ are generated by an AR(1) model as below. We first generate $\bar{h}(t)$ by

$$\bar{h}(t) = \kappa \times \bar{h}(t-1) + u_g(t), \tag{221}$$

where $u_g(t)$ are i.i.d. complex Gaussian variables with zero mean and unity variance per dimension. Then, we normalize $\bar{h}(t)$ and obtain $h(t)$ by

$$h(t) = \bar{h}(t) \times \sqrt{\frac{1-\kappa^2}{2}}. \tag{222}$$

The Rayleigh fading channel gains are generated using the parameters described in Table 13.

8.5.3.1 Performance of Optimal Mode Selection

In this simulation, we evaluate the performance of different optimal mode selection schemes, namely, Encoder A, B, and C. For Encoder B, we set MER=1% for optimal mode selection. The round-trip time is set to 10 ms. The delay-constrained hybrid ARQ uses the rates of Set $\mathcal{A}$ for RCPC codes and the delay bound is set to be 100 ms. Figure 86 shows the average PSNR-Y under a Rayleigh fading channel with different average SNR's and different Doppler rates. Again, we observe that Encoder C achieves the best performance, Encoder B has the second best performance, and Encoder A performs the worst. This demonstrates that our approach achieves better performance than the classical one, for video transmission over a Rayleigh fading channel. Furthermore, the three figures in Fig. 86 show that as the Doppler rate increases, the average PSNR-Y achieved also increases. This indicates that our hybrid ARQ can utilize the time diversity in the underlying fading channel to improve video presentation quality. This is expected since the delay bound (*i.e.*, 100 ms) is larger than the coherence time T_c of 33.8, 2.6, and 0.8 ms (see Table 13). That is, our hybrid ARQ can utilize the time diversity of the underlying fading channel, if the delay constraint imposed by the video application is larger than the coherence time of the fading channel.

8.5.3.2 Performance of Hybrid ARQ

To compare the performance of the truncated type-II hybrid ARQ and our delay-constrained hybrid ARQ, we again run our simulations under two scenarios where only Encoder C is employed.

In the first scenario, we set RTT=10 ms and the maximum number of retransmissions $N_m=3$ for the truncated hybrid ARQ. The truncated hybrid ARQ uses the rates of Set $\mathcal{B}$ for RCPC codes; our delay-constrained hybrid ARQ uses the rates of Set $\mathcal{A}$ for RCPC codes and the delay bound is set to be 100 ms. Figure 87(a) shows that our delay-constrained hybrid ARQ achieves higher average PSNR-Y than the truncated hybrid ARQ, for different average channel SNR and different Doppler rates. As we mentioned before, the reason is that our delay-constrained hybrid ARQ has lower probability of packet corruption. Compared with Figure 84(a), we see that for the same average channel SNR, the average PSNR-Y achieved under the Rayleigh fading channel is lower than that under the AWGN channel. This is conceivable since compared to an AWGN channel, a Rayleigh fading channel may cause long bursts of errors, which are hard to recover.

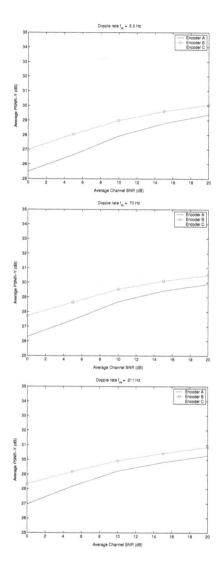

Figure 86: Performance of different video encoders under the setting of a Rayleigh fading channel.

235

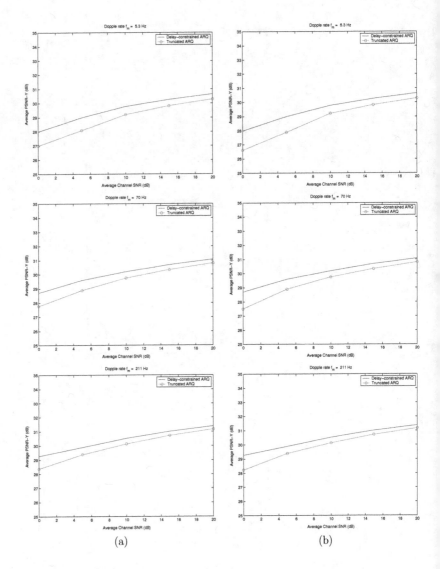

Figure 87: Average average PSNR-Y for hybrid ARQ schemes under the setting of a fading channel: (a) the first scenario, (b) the second scenario.

In the second scenario, we set RTT=30 ms and the maximum number of retransmissions N_m=10 for the truncated hybrid ARQ. Both hybrid ARQ scheme use the rates of Set $\mathcal{A}$ for RCPC codes. and the delay bound is set to be 100 ms. Figure 87(b) shows that our delay-constrained hybrid ARQ achieves higher average PSNR-Y than the truncated hybrid ARQ, for different average channel SNR and different Doppler rates. Again, the reason is that the truncated hybrid ARQ may waste bandwidth unnecessarily, which translates into lower average PSNR-Y.

In summary, the simulation results have demonstrated (1) the performance improvement of our approach over the classical approach for optimal mode selection, and (2) the performance improvement of our delay-constrained hybrid ARQ over the truncated Type-II hybrid ARQ, for real-time video transmission over wireless channels.

8.6 Summary

Due to high bit-error-rate in wireless environments, real-time video communication over wireless channels poses a challenging problem. To address this challenge, this chapter introduced adaptive QoS control to achieve robustness for video transmission over wireless. Our adaptive QoS control consists of optimal mode selection and delay-constrained hybrid ARQ. Our optimal mode selection provides QoS support (*i.e.*, error resilience) on the compression layer; it is adaptive to time-varying channel characteristics due to feedback of macroblock error ratio. Our delay-constrained hybrid ARQ provides QoS support (*i.e.*, reliability and bounded delay) on the link layer; it is adaptive to time-varying wireless channel due to its consideration of the display deadlines of the packets. The main contributions of this chapter are: (1) an optimal mode selection algorithm which provides the best trade-off between compression efficiency and error resilience in R-D sense, and (2) delay-constrained hybrid ARQ which is capable of providing reliability for the compression layer while guaranteeing delay bound and achieving high throughput. Simulation results have shown that the proposed adaptive QoS control outperforms the previous approaches for MPEG-4 video communications over wireless channels. The reason for this is that our adaptive QoS control combines the best features of error-resilient source encoding, FEC and delay-constrained retransmission.

CHAPTER 9

CONCLUSION

9.1 Summary of the Book

In this section, we summarize the research presented in this book.

We considered the problem of providing QoS guarantees in wireless networks. We addressed this problem primarily from the network perspective, but also studied it with the end-system-based approach. In Chapter 1, we introduced and motivated the problem. In Chapter 2, we overviewed issues in the area of network-centric QoS provisioning in wireless networks, which include network service models, traffic modeling, scheduling for wireless transmission, call admission control in wireless networks, and wireless channel modeling. In Chapter 3, we surveyed end-system-based QoS support techniques, including various congestion control and error control mechanisms.

In Chapter 4, we presented the effective capacity channel model. We modeled a wireless channel from the perspective of the communication *link-layer*. This is in contrast to existing channel models, which characterize the wireless channel at the *physical-layer*. Specifically, we modeled the wireless link in terms of two 'effective capacity' functions; namely, the probability of non-empty buffer $\gamma(\mu)$ and the QoS exponent $\theta(\mu)$. The QoS exponent is the inverse of a function which we call *effective capacity* (EC). The EC channel model is the dual of the effective bandwidth source traffic model, used in wired networks. Furthermore, we developed a simple and efficient algorithm to estimate the EC functions $\{\gamma(\mu), \theta(\mu)\}$. We provided key insights about the relations between the EC channel model and the physical-layer channel, *i.e.*, $\gamma(\mu)$ corresponds to the marginal CDF (*e.g.*, Rayleigh/Ricean distribution) while $\theta(\mu)$ is related to the Doppler spectrum. The EC channel model has been justified not only from a theoretical viewpoint (*i.e.*, Markov property of fading channels) but also from an experimental viewpoint (*i.e.*, the delay-violation probability does decay exponentially with the delay). We obtained link-layer QoS measures for various scenarios: flat-fading channels, frequency-selective fading channels, multi-link wireless networks, variable-bit-rate sources, packetized traffic, and wireless channels with non-negligible propagation delay. The QoS measures obtained can be easily translated into traffic envelope and service curve characterizations, which are popular in wired networks, such as ATM and IP, to provide guaranteed services. In addition, our channel model provides a general framework, under which physical-layer fading channels such as AWGN, Rayleigh fading, and Ricean fading channels can be studied.

In Chapters 5 and 6, we investigated the use of our EC channel model in designing admission control, resource reservation, and scheduling algorithms, for efficient support of QoS guarantees.

Chapter 5 addresses the problem of QoS provisioning for K users sharing a single time-slotted fading downlink channel. We developed simple and efficient schemes for admission control, resource allocation, and scheduling, to obtain a gain in delay-constrained capacity. Multiuser diversity obtained by the well-known K&H scheduling is the key that gives rise to this performance gain. However, the unique feature of this work is explicit support of a statistical QoS requirement, namely, a data rate, delay bound, and delay-bound violation probability triplet, for channels utilizing K&H scheduling. The concept of effective capacity is the key that explicitly guarantees the QoS. Thus, this work combines crucial ideas from the areas of communication theory and queueing theory to provide the tools to increase capacity and yet satisfy QoS constraints.

In Chapter 6, we examined the problem of providing QoS guarantees to K users over N parallel time-varying channels. We designed simple and efficient admission control, resource allocation, and scheduling algorithms for guaranteeing requested QoS. We developed two sets of scheduling algorithms, namely, joint K&H/RR scheduling and RC scheduling. The joint K&H/RR scheduling utilizes both multiuser diversity and frequency diversity to achieve capacity gain, and is an extension of Chapter 5. The RC scheduling is formulated as a linear program, which minimizes the channel usage while satisfying users' QoS constraints. The relation between the joint K&H/RR scheduling and the RC scheduling is that 1) if the admission control allocates channel resources to the RR scheduling due to tight delay requirements, then the RC scheduler can be used to minimize channel usage; 2) if the admission control allocates channel resources to the K&H scheduling only, due to loose delay requirements, then there is no need to use the RC scheduler. The key features of the RC scheduler are high efficiency, simplicity, and statistical QoS support.

In Chapter 7, we investigated practical aspects of the effective capacity approach. We showed how to quantify the effect of practical modulation and channel coding on the effective capacity approach to QoS provisioning. We identified the fundamental trade-off between power and time diversity in QoS provisioning over fading channels, and proposed a novel time-diversity dependent power control scheme to leverage time diversity. Compared to the power control used in the 3G wireless systems, our time-diversity dependent power control can achieve a factor of 2 to 4.7 capacity gain for mobile speeds between 0.11 m/s and 56 m/s. Equipped with the time-diversity dependent power control and the effective capacity approach, we designed a power control and scheduling mechanism, which is robust in QoS provisioning against large scale fading and non-stationary small scale fading. With this mechanism, we proposed QoS provisioning algorithms for downlink and uplink transmissions, respectively. We also considered QoS provisioning for multiple users.

In Chapter 8, we introduced adaptive QoS control to achieve robustness for real-time video communication over wireless channels. Our adaptive QoS control consists of optimal mode selection and delay-constrained hybrid ARQ. Our optimal mode selection provides QoS support (*i.e.*, error resilience) on the compression layer; it is adaptive to time-varying channel characteristics due to feedback of macroblock error ratio. Our delay-constrained hybrid ARQ provides QoS support (*i.e.*, reliability and bounded delay) on the link layer; it is adaptive to time-varying wireless channel due to its consideration of the display deadlines of the packets. The optimal mode selection algorithm provides the best trade-off between compression efficiency and error resilience in the rate-distortion sense, and the delay-constrained hybrid ARQ is capable of providing reliability for the compression layer while guaranteeing delay bound and achieving high throughput.

The theme of our research is supporting QoS in wireless networks, with simplicity, efficiency and practicality in mind. In conducting the research, we took both network-centric and end-system-based approach and demonstrated the advantages of our approaches through analyses and simulations.

The effective capacity model we proposed captures the effect of channel fading on the queueing behavior of the link, using a computationally simple yet accurate model, and thus, is the critical device we need to design efficient QoS provisioning mechanisms. We call such an approach to channel modeling and QoS provisioning, as *effective capacity approach*.

The effective capacity approach represents a paradigm fundamentally different from the conventional approaches to wireless networking, not only in channel modeling, but also in QoS provisioning. Specifically, we model the wireless channel at the link layer, in contrast to the conventional modeling at the physical layer; we use the EC channel model to relate the control parameters of our QoS provisioning mechanisms to connection-level QoS measures, in contrast to conventional schemes that are unable to directly relate the control parameters of the schemes to connection-level QoS measures. As a result, our QoS provisioning mechanisms can explicitly provide QoS guarantees, in contrast to conventional schemes that only provide implicit QoS.

With the effective capacity approach, we are able to design schemes that efficiently and explicitly provide statistical QoS guarantees in wireless networks as shown in Chapters 5 and 6. We also have studied the viability of the EC approach in practical situations in Chapter 7.

Like everything else, the effective capacity approach also has its limitations. First, it requires estimation of channel gains at the receivers, and hence the system performance depends on the accuracy of the estimation. Second, the EC approach requires feedback of channel estimates, which incurs signaling overhead; in addition, the BER condition and the propagation delay of the feedback channel will degrade the system performance. Third, the EC approach is not suitable for AWGN channels or fading channels with low degree of time diversity. Fourth, non-stationarity of the channel gain process, which is typical in practical situations, makes the estimate of the QoS exponent $\theta(\mu)$ inaccurate. Last but not least, since the EC approach follows the general idea that modeling at a higher layer hides the complexity of the lower layers, the effect of changes in the lower layers on the effective capacity is not directly obvious.

In summary, this research introduced a novel effective-capacity approach to modeling fading channels and provisioning QoS. Although there are still many issues to be resolved, we believe that our EC channel model, which was specifically constructed keeping in mind the link-layer QoS measures, will find wide applicability in future wireless networks that need QoS provisioning.

9.2 Future Work

In this section, we point out future research directions.

Optimization of the Physical Layer Design with the Effective Capacity approach

In this book, we introduced the notion of effective capacity or statistical delay-constrained capacity of a fading channel (*i.e.*, the maximum data rate achievable with the delay-bound violation probability satisfied). The effective capacity can be used to formulate optimization problems for physical-layer designs, *e.g.*, space-time coding, orthogonal frequency division multiplexing (OFDM), LDPC codes, multiple input multiple output (MIMO) systems, multiuser detection, equalization, etc. For example, one problem is to choose one physical-layer design among multiple candidate designs so as to maximize the effective capacity. Another problem is, for a given physical layer design, how to minimize the transmit power or channel usage subject to the link-layer QoS constraints; the effective capacity serves as a tool to connect the physical layer design to the link-layer performance. Hence, based on the effective capacity approach, the physical-layer design may require a substantial change. An example is our time-diversity dependent power control proposed in Chapter 7.

End-to-end Support for Statistical QoS Guarantees

The effective capacity approach and QoS measures developed in Chapter 4, can be used to design efficient mechanisms to provide end-to-end QoS guarantees in a multihop wireless network. This will involve developing algorithms for QoS routing, resource reservation, admission control and scheduling. The problem of QoS routing can be formulated as below: given a network and the effective capacity function of each link, find all feasible paths that satisfy the requested link-layer QoS, and then select the least-cost path(s) for resource reservation. The effective capacity function is used to connect the path performance to the QoS requirements requested by the users. The resource reservation and admission control can be designed, based on the effective capacity functions of the channels in the network. Since the existing QoS routing schemes for *ad hoc* networks are targeted at providing *deterministic* QoS, routing for *statistical* QoS provisioning in *ad hoc* networks has not been studied, to the best of our knowledge. We believe that the effective capacity approach will provide a powerful tool in designing new routing schemes for statistical QoS provisioning in *ad hoc* networks.

Joint Optimization for Video Communication over Wireless Channels: a Multi-layer Approach

In Ref. [149], we presented an *end-to-end approach* to generalize the classical theory of R-D optimized mode selection for point-to-point video communication, and introduced a notion of *global* (end-to-end) distortion by taking into consideration both the path characteristics (i.e., packet loss) and the receiver behavior (i.e., the error concealment scheme), in addition to the source behavior (i.e., quantization distortion and packetization). In Chapter 8, we applied this technique to wireless video transmission. The global distortion can be used to formulate cross-layer optimization problems. For example, one problem is to minimize the transmit power, subject to a constant global distortion constraint; another problem is to minimize the global distortion, subject to average/peak power constraint. The global distortion could be meant for a macroblock, a group of blocks (GOBs), or an image frame. The global distortion is a function of the bit error rate; the bit error rate depends on the transmit power, the modulation scheme, channel coding gain, and the transmission rate. Hence, to solve the cross-layer optimization problems for video transmission over fading channels, one may use 1) adaptive power control, 2) adaptive modulation and channel coding, 3) adaptive retransmission, and 4) adaptive source rate control. We believe that such a joint optimization can achieve better performance than designing each layer separately.

APPENDIX

A.1 A Fast Binary Search Procedure for Channel Estimation

We show that the $\{\gamma^{(c)}(\mu), \theta^{(c)}(\mu)\}$ functions that specify our effective capacity channel model, can be easily used to obtain the service curve specification $\Psi(t) = \{\sigma^{(c)}, \lambda_s^{(c)}\}$. The parameter $\sigma^{(c)}$ is simply equal to the source delay requirement D_{max}. Thus, only the channel sustainable rate $\lambda_s^{(c)}$ needs to be estimated. $\lambda_s^{(c)}$ is the source rate μ at which the required QoS (delay-violation probability ε) is achieved.

The following binary search procedure estimates $\lambda_s^{(c)}$ for a given (unknown) fading channel and source specification $\{D_{max}, \varepsilon\}$. In the algorithm, ϵ_e is the error between the target and the estimated $\sup_t Pr\{D(t) \geq D_{max}\}$, ϵ_t the precision tolerance, μ the source rate, μ_l a lower bound on the source rate, and μ_u an upper bound on the source rate.

Algorithm A.1 (Estimation of the channel sustainable rate $\lambda_s^{(c)}$)

/ Initialization */*

Initialize ε, ϵ_t, and ϵ_e;

/ E.g., $\varepsilon = 10^{-3}$; $\epsilon_t = 10^{-2}$; $\epsilon_e = 1$; */*

$\mu_l := 0$; / An obvious lower bound on the rate */*

$\mu_u := r_{awgn}$; / An obvious upper bound on the rate is the AWGN capacity */*

$\mu := (\mu_l + \mu_u)/2$;

/ Binary search to find a $\lambda_s^{(c)}$ that is conservative and within ϵ_t */*

While $((\epsilon_e/\varepsilon > \epsilon_t)$ or $(\epsilon_e < 0))$ {

    The data source transmits at the output rate μ;

    Estimate $\hat{\gamma}$ and $\hat{\theta}$ using (36) to (39);

    Use Eq. (40) to obtain $\sup_t Pr\{D(t) \geq D_{max}\}$;

    $\epsilon_e := \varepsilon - \sup_t Pr\{D(t) \geq D_{max}\}$;

```
if (ε_e ≥ 0) { /* Conservative */

    if (ε_e/ε > ε_t) { /* Too conservative */

        μ_l := μ; /* Increase rate */

        μ := (μ_l + μ_u)/2;

    }

}

else { /* Optimistic */

    μ_u := μ; /* Reduce rate */

    μ := (μ_l + μ_u)/2;

}

}
```

$\lambda_s^{(c)} := \mu.$

Algorithm A.1 uses a binary search to find the channel sustainable rate $\lambda_s^{(c)}$. An alternative approach is to use a parallel search, such as the one described in Ref. [88]. A parallel search would require more computations, but would converge faster.

A.2 Proof of Proposition 4.1

Denote $r_i(t)$ the instantaneous channel capacity of sub-channel i ($i = 1, \cdots, N$) of the frequency-selective fading channel, at time t. Then for $u \geq 0$, we have

$$
\begin{aligned}
\alpha_N(u) &\overset{(a)}{=} -\lim_{t\to\infty} \frac{1}{ut} \log E[e^{-u\int_0^t r_N(\tau)\mathrm{d}\tau}] \\
&\overset{(b)}{=} -\lim_{t\to\infty} \frac{1}{ut} \log E[e^{-u\int_0^t \frac{1}{N}\sum_{i=1}^N r_i(\tau)\mathrm{d}\tau}] \\
&\overset{(c)}{=} -\lim_{t\to\infty} \frac{1}{ut} \log(E[e^{-u\int_0^t \frac{1}{N}r_1(\tau)\mathrm{d}\tau}])^N \\
&= -\lim_{t\to\infty} \frac{1}{t\frac{u}{N}} \log E[e^{-\frac{u}{N}\int_0^t r_1(\tau)\mathrm{d}\tau}] \\
&\overset{(d)}{=} \alpha_1(\frac{u}{N})
\end{aligned}
\tag{223}
$$

where (a) from (75), (b) since $r_N(t) = \frac{1}{N}\sum_{i=1}^{N} r_i(t)$, (c) since $r_i(t)$ ($i = 1, \cdots, N$) are i.i.d., and (d) from (76). Then by $\alpha_N(u) = \alpha_1(\frac{u}{N}) \doteq \mu$, we have

$$u = \alpha_N^{-1}(\mu) \tag{224}$$

and

$$\frac{u}{N} = \alpha_1^{-1}(\mu). \tag{225}$$

Removing u in (224) and (225) results in

$$\alpha_N^{-1}(\mu) = N \times \alpha_1^{-1}(\mu) \tag{226}$$

Thus, we have

$$
\begin{aligned}
\theta_N^{(c)}(\mu) &\overset{(a)}{=} \mu \alpha_N^{-1}(\mu) \\
&\overset{(b)}{=} \mu \times N \times \alpha_1^{-1}(\mu) \\
&\overset{(c)}{=} N \times \theta_1^{(c)}(\mu)
\end{aligned}
\tag{227}
$$

where (a) from (77), (b) from (226), and (c) from (78). This completes the proof. ∎

A.3 Proof of Proposition 4.2

Denote $Q_k(t)$ the queue length at time t at node k ($k = 1, \cdots, K$), $Q(t)$ the end-to-end queue length at time t, $Q(\infty)$ the steady state of the end-to-end queue length, $A(t_0, t)$ the amount of arrival to node 1 (see Figure 40) over the time interval $[t_0, t]$. Define $\tilde{S}_k(t_0, t) = \int_{t_0}^{t} r_k(\tau)\mathrm{d}\tau$, which is the service provided by channel k over the time interval $[t_0, t]$.

We first prove an upper bound. It can be proved [167, page 81] that

$$Q(t) = \sum_{k=1}^{K} Q_k(t) = \sup_{0 \leq t_0 \leq t} \left\{ A(t_0, t) - \tilde{S}(t_0, t) \right\} \tag{228}$$

where $\tilde{S}(t_0, t)$ is defined by (85). Without loss of generality, we consider the discrete time case only, i.e., $t \in \mathbb{N}$, where $\mathbb{N}$ is the set of natural numbers. From (228) and Loynes Theorem [85], we obtain

$$Q(\infty) = \sup_{t \in \mathbb{N}} \left\{ A(0, t) - \tilde{S}(0, t) \right\} = \sup_{t \in \mathbb{N}} \left\{ \mu t - \tilde{S}(0, t) \right\} \tag{229}$$

245

Then, we have

$$Pr\{Q(\infty) > q\} = Pr\left\{\sup_{t\in\mathbb{N}}\left\{\mu t - \tilde{S}(0,t)\right\} > q\right\} \tag{230}$$

$$\overset{(a)}{\leq} Pr\left\{\bigcup_{t\in\mathbb{N}}\left\{\mu t - \tilde{S}(0,t) > q\right\}\right\} \tag{231}$$

$$\overset{(b)}{\leq} \sum_{t\in\mathbb{N}} Pr\left\{\mu t - \tilde{S}(0,t) > q\right\} \tag{232}$$

$$\overset{(c)}{\leq} \sum_{t\in\mathbb{N}} e^{-uq} E[e^{u(\mu t - \tilde{S}(0,t))}] \tag{233}$$

where (a) since the event $\left\{\sup_{t\in\mathbb{N}}\left\{\mu t - \tilde{S}(0,t)\right\} > q\right\} \subset \bigcup_{t\in\mathbb{N}}\left\{\mu t - \tilde{S}(0,t) > q\right\}$, (b) is due to the union bound, and (c) from the Chernoff bound. Since $\alpha(u) = -\Lambda_{tandem}(-u)/u$, we have

$$\alpha(u) = -\lim_{t\to\infty}\frac{1}{ut}\log E[e^{-u\tilde{S}(0,t)}], \ \forall \ u > 0, \tag{234}$$

Hence, for any $\epsilon > 0$, there exists a number $\tilde{t} > 0$ such that for $t \geq \tilde{t}$, we have

$$E[e^{-u\tilde{S}(0,t)}] \leq e^{u(-\alpha(u)+\epsilon)t}, \qquad \forall u > 0. \tag{235}$$

If $\mu + \epsilon < \alpha(u)$, we have

$$\sum_{t\in\mathbb{N}} e^{-uq} E[e^{u(\mu t - \tilde{S}(0,t))}] \overset{(a)}{\leq} \sum_{t\geq\tilde{t}} e^{-uq} e^{u(\mu-\alpha(u)+\epsilon)t} + \sum_{t=1}^{\tilde{t}-1} e^{-uq} E[e^{u(\mu t - \tilde{S}(0,t))}] \tag{236}$$

$$\overset{(b)}{\leq} e^{-uq} \times \left(\frac{e^{u(\mu-\alpha(u)+\epsilon)\tilde{t}}}{1 - e^{u(\mu-\alpha(u)+\epsilon)}} + \sum_{t=1}^{\tilde{t}-1} E[e^{u(\mu t - \tilde{S}(0,t))}]\right) \tag{237}$$

where (a) from (235), and (b) from geometric sum. From (233) and (237), we have

$$Pr\{Q(\infty) > q\} \leq \gamma \times e^{-uq}, \qquad \text{if } \mu + \epsilon < \alpha(u), \tag{238}$$

where γ is a constant independent of q. Hence, we obtain

$$\limsup_{q\to\infty}\frac{1}{q}\log Pr\{Q(\infty) > q\} \leq -u, \qquad \text{if } \mu + \epsilon < \alpha(u). \tag{239}$$

246

Letting $\epsilon \to 0$, we have

$$\limsup_{q \to \infty} \frac{1}{q} \log Pr\left\{Q(\infty) > q\right\} \leq -u, \qquad \text{if } \mu < \alpha(u). \tag{240}$$

Since $Q(t) = \mu \times D(t)$, $\forall t \geq 0$, (240) results in

$$\limsup_{D_{max} \to \infty} \frac{1}{D_{max}} \log Pr\{D(\infty) > D_{max}\} \leq -\mu \times u, \qquad \text{if } \mu < \alpha(u). \tag{241}$$

Let $\theta = \mu \times u$. Then (241) becomes (88).

Next we prove a lower bound. Let $q = \beta t$ where $\beta > 0$. Then, we have

$$\liminf_{q \to \infty} \frac{1}{q} \log Pr\left\{Q(\infty) > q\right\} = \liminf_{t \to \infty} \frac{1}{\beta t} \log Pr\left\{Q(\infty) > \beta t\right\} \tag{242}$$

$$= \liminf_{t \to \infty} \frac{1}{\beta t} \log Pr\left\{\sup_{t \in \mathbb{N}}\left\{\mu t - \tilde{S}(0,t)\right\} > \beta t\right\} \tag{243}$$

$$\geq \liminf_{t \to \infty} \frac{1}{\beta t} \log Pr\left\{\mu t - \tilde{S}(0,t) > \beta t\right\} \tag{244}$$

$$= \liminf_{t \to \infty} \frac{1}{\beta t} \log Pr\left\{\frac{-\tilde{S}(0,t)}{t} > \beta - \mu\right\} \tag{245}$$

$$\overset{(a)}{\geq} -\frac{1}{\beta} \inf_{x > \beta - \mu} \Lambda^*(x) \tag{246}$$

where (a) from Gärtner-Ellis Theorem [24] since $\Lambda_{tandem}(u)$ satisfies Property 4.1, and the Legendre-Fenchel transform $\Lambda^*(x)$ of $\Lambda_{tandem}(-u)$ is defined by

$$\Lambda^*(x) = \sup_{u \in \mathbb{R}}\{u \times x - \Lambda_{tandem}(-u)\}. \tag{247}$$

Since (246) holds for any $\beta > 0$, we have

$$\liminf_{q \to \infty} \frac{1}{q} \log Pr\left\{Q(\infty) > q\right\} \geq \sup_{\beta > 0} -\frac{1}{\beta} \inf_{x > \beta - \mu} \Lambda^*(x) \tag{248}$$

$$= -\inf_{y > -\mu} \frac{\Lambda^*(y)}{y + \mu} \tag{249}$$

It can be proved [38] that

$$\inf_{y > -\mu} \frac{\Lambda^*(y)}{y + \mu} = u^*, \qquad \text{where } \Lambda_{tandem}(-u^*) = -\mu \times u^*. \tag{250}$$

247

From the definition of effective capacity $\alpha(u)$ in (234), $\Lambda_{tandem}(-u^*) = -\mu \times u^*$ implies $\alpha(u^*) = \mu$. Then, applying (250) to (249) leads to

$$\liminf_{q \to \infty} \frac{1}{q} \log Pr\left\{Q(\infty) > q\right\} \geq u^*, \qquad \text{where } \alpha(u^*) = \mu. \qquad (251)$$

Due to the continuity of $\alpha(u)$, we have

$$\lim_{u \to u^*} \alpha(u) = \mu \qquad (252)$$

Hence, letting $u \to u^*$, (240) and (251) result in

$$u^* \leq \liminf_{q \to \infty} \frac{1}{q} \log Pr\left\{Q(\infty) > q\right\} \leq \limsup_{q \to \infty} \frac{1}{q} \log Pr\left\{Q(\infty) > q\right\} \leq u^* \quad (253)$$

Hence, we have

$$\lim_{q \to \infty} \frac{1}{q} \log Pr\left\{Q(\infty) > q\right\} = u^*, \qquad \text{where } \alpha(u^*) = \mu. \qquad (254)$$

Since $Q(t) = \mu \times D(t)$, $\forall t \geq 0$, (254) results in

$$\lim_{D_{max} \to \infty} \frac{1}{D_{max}} \log Pr\{D(\infty) > D_{max}\} = -\mu \times u^*, \qquad \text{where } \alpha(u^*) = \mu. \qquad (255)$$

Let $\theta^* = \mu \times u^*$. Then (255) becomes (89).

From (85), it is obvious that

$$\tilde{S}(t_0, t) \leq \min_k \tilde{S}_k(t_0, t). \qquad (256)$$

Then we have for $u > 0$,

$$\alpha(u) = -\lim_{t \to \infty} \frac{1}{ut} \log E[e^{-u\tilde{S}(0,t)}] \qquad (257)$$

$$\overset{(a)}{\leq} -\lim_{t \to \infty} \frac{1}{ut} \log E[e^{-u \min_k \tilde{S}_k(0,t)}] \qquad (258)$$

$$\leq \min_k -\lim_{t \to \infty} \frac{1}{ut} \log E[e^{-u\tilde{S}_k(0,t)}] \qquad (259)$$

$$= \min_k \alpha_k(u) \qquad (260)$$

where (a) from (256). This completes the proof. ∎

248

A.4 Proof of Proposition 4.3

Denote $r_k(t)$ $(k = 1, \cdots, K)$ channel capacity of link k at time t. From Figure 41, it is clear that the network has only one queue and multiple servers, each of which corresponds to a wireless link. Since the total instantaneous channel capacity $r(t) = \sum_{k=1}^{K} r_k(t)$, the effective capacity function for the aggregate parallel links is

$$
\begin{aligned}
\alpha(u) &\overset{(a)}{=} -\lim_{t \to \infty} \frac{1}{ut} \log E[e^{-u \int_0^t r(\tau) d\tau}] \\
&= -\lim_{t \to \infty} \frac{1}{ut} \log E[e^{-u \int_0^t \sum_{k=1}^K r_k(\tau) d\tau}] \\
&\overset{(b)}{=} -\lim_{t \to \infty} \frac{1}{ut} \sum_{k=1}^{K} \log E[e^{-u \int_0^t r_k(\tau) d\tau}] \\
&\overset{(c)}{=} \sum_{k=1}^{K} \alpha_k(u)
\end{aligned}
\tag{261}
$$

where (a) from (28), (b) since $\{r_k(t), k = 1, \cdots, K\}$ are independent, and (c) from (87).

Given the effective capacity $\alpha(u)$, we can prove (91) and (92) with the same technique used in proving (88) and (89). This completes the proof. ∎

A.5 Proof of Proposition 4.4

For the traffic of constant rate $\lambda_s^{(s)}$, denote $\tilde{Q}(\infty)$ the steady state of the end-to-end queue length and $\tilde{D}(\infty)$ the end-to-end delay. Using the result in [76, page 30]), we can show

$$
D(\infty) - \tilde{D}(\infty) \le \sigma^{(s)}/\lambda_s^{(s)},
\tag{262}
$$

Note that $D(\infty)$ is the end-to-end delay experienced by the traffic constrained by a leaky bucket with bucket size $\sigma^{(s)}$ and token generating rate $\lambda_s^{(s)}$. Hence, we have

$$
D(\infty) \le \tilde{D}(\infty) + \sigma^{(s)}/\lambda_s^{(s)},
\tag{263}
$$

From (238) and $\tilde{Q}(\infty) = \lambda_s^{(s)} \times \tilde{D}(\infty)$, we have

$$
Pr\left\{\tilde{D}(\infty) > D_{max}\right\} \le \gamma \times e^{-u \times \lambda_s^{(s)} \times D_{max}}, \qquad \text{if } \alpha(u) > \lambda_s^{(s)},
\tag{264}
$$

where γ is a constant independent of D_{max}. Then, we have

$$Pr\{D(\infty) > D_{max}\} \overset{(a)}{=} Pr\left\{\tilde{D}(\infty) > D_{max} - \sigma^{(s)}/\lambda_s^{(s)}\right\}$$

$$\overset{(b)}{\leq} \gamma \times e^{-u \times \lambda_s^{(s)} \times (D_{max} - \sigma^{(s)}/\lambda_s^{(s)})} \tag{265}$$

where (a) from (263), and (b) from (264). Hence, we have

$$\limsup_{D_{max} \to \infty} \frac{1}{D_{max} - \sigma^{(s)}/\lambda_s^{(s)}} \log Pr\{D(\infty) > D_{max}\} \leq -\theta, \qquad \text{if } \alpha(\theta/\lambda_s^{(s)}) > \lambda_s^{(s)}.$$
$$\tag{266}$$

Similar to the proof of Proposition 4.2, we can obtain a lower bound

$$\liminf_{D_{max} \to \infty} \frac{1}{D_{max} - \sigma^{(s)}/\lambda_s^{(s)}} \log Pr\{D(\infty) > D_{max}\} \geq -\theta^*, \qquad \text{where } \alpha(\theta^*/\lambda_s^{(s)}) = \lambda_s^{(s)}.$$
$$\tag{267}$$

Combining (266) and (267), we obtain (94). This completes the proof. ∎

A.6 Proof of Proposition 4.5

The proof is similar to that of Proposition 4.2.

Denote $Q(t)$ the end-to-end queue length at time t, $Q(\infty)$ the steady state of the end-to-end queue length, $A(t_0, t)$ the amount of external arrival to the network over the time interval $[t_0, t]$. From (228), we know

$$Q(t) = \sup_{0 \leq t_0 \leq t} \left\{A(t_0, t) - \tilde{S}(t_0, t)\right\} \tag{268}$$

where $\tilde{S}(t_0, t)$ is defined by (85) for the tandem links, and is defined by $\tilde{S}(t_0, t) = \int_0^t \sum_{k=1}^{K} r_k(\tau)d\tau$ for independent parallel links.

We first prove an upper bound. Without loss of generality, we consider the discrete time case only, $i.e.$, $t \in \mathbb{N}$, where $\mathbb{N}$ is the set of natural numbers. From (268) and Loynes Theorem [85], we obtain

$$Q(\infty) = \sup_{t \in \mathbb{N}} \left\{A(0, t) - \tilde{S}(0, t)\right\} \tag{269}$$

Then, we have

$$Pr\left\{Q(\infty) > q\right\} \;=\; Pr\left\{\sup_{t\in\mathbb{N}}\left\{A(0,t) - \tilde{S}(0,t)\right\} > q\right\} \tag{270}$$

$$\leq\; Pr\left\{\bigcup_{t\in\mathbb{N}}\left\{A(0,t) - \tilde{S}(0,t) > q\right\}\right\} \tag{271}$$

$$\leq\; \sum_{t\in\mathbb{N}} Pr\left\{A(0,t) - \tilde{S}(0,t) > q\right\} \tag{272}$$

$$\leq\; \sum_{t\in\mathbb{N}} e^{-uq} E\big[e^{u(A(0,t)-\tilde{S}(0,t))}\big] \tag{273}$$

From the definition of effective capacity in (234), for any $\epsilon/2 > 0$, there exists a number $\tilde{t} > 0$ such that for $t \geq \tilde{t}$, we have

$$E\big[e^{-u\tilde{S}(0,t)}\big] \leq e^{u(-\alpha(u)+\epsilon/2)t}, \qquad \forall u > 0. \tag{274}$$

Similarly, from the definition of effective bandwidth in (22), for any $\epsilon/2 > 0$, there exists a number $\tilde{t} > 0$ such that for $t \geq \tilde{t}$, we have

$$E\big[e^{uA(0,t)}\big] \leq e^{u(\alpha^{(s)}(u)+\epsilon/2)t}, \qquad \forall u > 0. \tag{275}$$

Without loss of generality, here we choose the same $\tilde{t}$ for both (274) and (275), since we can always choose the maximum of the two to make (274) and (275) hold. Then, if $\alpha^{(s)}(u) + \epsilon < \alpha(u)$, we have

$$\sum_{t\in\mathbb{N}} e^{-uq} E\big[e^{u(A(0,t)-\tilde{S}(0,t))}\big] \;\overset{(a)}{\leq}\; \sum_{t\geq\tilde{t}} e^{-uq} e^{u(\alpha^{(s)}(u)-\alpha(u)+\epsilon)t} + \sum_{t=1}^{\tilde{t}-1} e^{-uq} E\big[e^{u(A(0,t)-\tilde{S}(0,t))}\big] \tag{276}$$

$$\leq\; e^{-uq} \times \left(\frac{e^{u(\alpha^{(s)}(u)-\alpha(u)+\epsilon)\tilde{t}}}{1 - e^{u(\alpha^{(s)}(u)-\alpha(u)+\epsilon)}} + \sum_{t=1}^{\tilde{t}-1} E\big[e^{u(A(0,t)-\tilde{S}(0,t))}\big]\right) \tag{277}$$

where (a) from (274) and (275). From (273) and (277), we have

$$\limsup_{q\to\infty} \frac{1}{q}\log Pr\left\{Q(\infty) > q\right\} \;\leq\; -u, \text{ if } \alpha^{(s)}(u) + \epsilon < \alpha(u). \tag{278}$$

Letting $\epsilon \to 0$, we have

$$\limsup_{q\to\infty} \frac{1}{q}\log Pr\left\{Q(\infty) > q\right\} \;\leq\; -u, \text{ if } \alpha^{(s)}(u) < \alpha(u). \tag{279}$$

251

Similar to the proof of Proposition 4.2, we can obtain a lower bound

$$\liminf_{q \to \infty} \frac{1}{q} \log Pr\{Q(\infty) > q\} \geq -u^*, \text{ where } \alpha^{(s)}(u^*) = \alpha(u^*). \quad (280)$$

Combining (279) and (280), we have

$$\lim_{q \to \infty} \frac{1}{q} \log Pr\{Q(\infty) > q\} = -u^*, \text{ where } \alpha^{(s)}(u^*) = \alpha(u^*). \quad (281)$$

Since $Q(\infty) = \alpha^{(s)}(u^*) \times D(\infty)$, (281) results in (97). This completes the proof. ∎

A.7 Proof of Proposition 4.6

For the packetized traffic, denote $Q_k(t)$ the queue length at time t at node k ($k = 1, \cdots, N$), $Q(t)$ the end-to-end queue length at time t, and $Q(\infty)$ the steady state of the end-to-end queue length. Correspondingly, for the 'fluid' traffic of constant arrival rate μ, denote $\tilde{Q}_k(t)$ the queue length at time t at node k, $\tilde{Q}(t)$ the end-to-end queue length at time t, $\tilde{Q}(\infty)$ the steady state of the end-to-end queue length, and $\tilde{D}(\infty)$ the end-to-end delay.

For each node k, we have the sample path relation as below [99]

$$Q_k(t) - \tilde{Q}_k(t) \leq L_c, \qquad \forall t \geq 0. \quad (282)$$

Summing up over k, we obtain

$$\sum_{k=1}^{N} [Q_k(t) - \tilde{Q}_k(t)] = Q(t) - \tilde{Q}(t) \leq N \times L_c, \qquad \forall t \geq 0. \quad (283)$$

Hence, for the steady state, we have

$$Q(\infty) - \tilde{Q}(\infty) \leq N \times L_c. \quad (284)$$

Since $Q(\infty) = \mu \times D(\infty)$ and $\tilde{Q}(\infty) = \mu \times \tilde{D}(\infty)$, we have

$$D(\infty) - \tilde{D}(\infty) \leq N \times L_c/\mu. \quad (285)$$

Note that $D(\infty)$ is the end-to-end delay experienced by the packetized traffic with constant bit rate μ and constant packet size L_c. Then, we can prove (98) in the same way as we prove (94) in Proposition 4.4. ∎

A.8   Proof of Proposition 4.7

Denote $\tilde{D}(\infty)$ the end-to-end delay experienced by the 'fluid' traffic with constant arrival rate $\lambda_s^{(s)}$. Using the sample path relation in [76, page 35]), we obtain

$$D(\infty) - \tilde{D}(\infty) \leq N \times L_{max}/\lambda_s^{(s)} + \sigma^{(s)}/\lambda_s^{(s)}, \qquad (286)$$

Note that $D(\infty)$ is the end-to-end delay experienced by the packetized traffic having maximum packet size L_{max} and constrained by a leaky bucket with bucket size $\sigma^{(s)}$ and token generating rate $\lambda_s^{(s)}$. Then, we can prove (100) in the same way as we prove (94) in Proposition 4.4. ∎

A.9   Proof of Proposition 4.8

Denote $\tilde{D}(\infty)$ the end-to-end delay experienced by the fluid traffic with constant arrival rate μ and without propagation delay. Using the sample path relation between the two cases (with/without propagation delay), it is easy to show

$$D(\infty) - \tilde{D}(\infty) \leq \sum_{i=1}^{N} d_i, \qquad (287)$$

Then, we can prove (102) in the same way as we prove (94) in Proposition 4.4. ∎

A.10   Proof of Proposition 4.9

Denote $\tilde{D}(\infty)$ the end-to-end delay experienced by the 'fluid' traffic with constant arrival rate $\lambda_s^{(s)}$ and without propagation delay. Using the sample path relation in [76, page 35]), we obtain

$$D(\infty) - \tilde{D}(\infty) \leq N \times L_{max}/\lambda_s^{(s)} + \sigma^{(s)}/\lambda_s^{(s)} + \sum_{i=1}^{N} d_i, \qquad (288)$$

Then, we can prove (104) in the same way as we prove (94) in Proposition 4.4. ∎

A.11   Proof of Proposition 5.1

In this proof, the time index t for channel gains and capacities is dropped, due to the assumption of stationarity of the channel gains. Without loss of generality, the expectation of channel power gain g_k can be normalized to one, *i.e.*, $\mathbf{E}[g_k] = 1, \forall k$. Then, for Rayleigh fading channels, the CDF of each channel power gain g_k is

$$F_G(g) = 1 - e^{-g}. \tag{289}$$

For low SNR, *i.e.*, very small g_k, the channel capacity of the kth user can be approximated by

$$c_k = \log_2(1 + g_k) \approx g_k. \tag{290}$$

Then we have

$$
\begin{aligned}
c_{max}/\mathbf{E}[c_1] &= \mathbf{E}[\max[c_1, c_2, \cdots, c_K]]/\mathbf{E}[c_1] && (291) \\
&\approx \mathbf{E}[\max[g_1, g_2, \cdots, g_K]]/\mathbf{E}[g_1] && (292) \\
&\stackrel{(a)}{=} \int_0^\infty g \mathrm{d}(F_G^K(g)) && (293) \\
&= \int_0^\infty (1 - F_G^K(g))\mathrm{d}g && (294) \\
&= \int_0^\infty (1 - F_G(g))\frac{(1 - F_G^K(g))}{(1 - F_G(g))}\mathrm{d}g && (295) \\
&= \int_0^\infty e^{-g}(F_G^{K-1}(g) + F_G^{K-2}(g) + \cdots + 1)\mathrm{d}g && (296) \\
&= \int_0^\infty (F_G^{K-1}(g) + F_G^{K-2}(g) + \cdots + 1)\mathrm{d}F_G(g) && (297) \\
&= 1 + \frac{1}{2} + \cdots + \frac{1}{K} && (298)
\end{aligned}
$$

where (a) follows from the fact that the CDF of the random variable $\max[g_1, g_2, \cdots, g_K]$ is $F_G^K(g)$ [112, page 248], and $\mathbf{E}[g_1] = 1$. It can be easily proved that

$$\log(K + 1) \leq \sum_{k=1}^K \frac{1}{k} \leq 1 + \log K. \tag{299}$$

So for large K, we can approximate $\sum_{k=1}^K \frac{1}{k}$ by $\log(K + 1)$. This completes the proof. ∎

A.12 Proof of Proposition 5.2

We use the notation in Section 5.3.1.2. From (28), we have

$$\alpha_{K,\varsigma,\beta}(u) = -\lim_{t \to \infty} \frac{1}{ut} \log E[e^{-u \int_0^t r(\tau)\mathrm{d}\tau}], \ \forall \ u \geq 0. \tag{300}$$

In Chapter 5, since t is a discrete frame index, the integral above should be thought of as a summation.

Since we only consider the homogeneous case, without loss of generality, denote $\alpha_{K,\lambda\zeta,\lambda\beta}(u)$ the effective capacity function of user $k = 1$ under K&H/RR scheduling, with frame shares $\lambda\zeta$ and $\lambda\beta$ respectively ($\lambda > 0$). Denote the resulting capacity process allotted to user 1 by the joint scheduler as the process $\hat{r}(t)$. Then for $u \geq 0$, we have

$$
\begin{aligned}
\alpha_{K,\lambda\zeta,\lambda\beta}(u) \overset{(a)}{=}\;& \frac{-\lim_{t\to\infty}\frac{1}{t}\log E[e^{-u\int_0^t \hat{r}(\tau)\mathrm{d}\tau}]}{u} \\
\overset{(b)}{=}\;& \frac{-\lim_{t\to\infty}\frac{1}{t}\log E[e^{-u\int_0^t \lambda r(\tau)\mathrm{d}\tau}]}{u} \\
=\;& \frac{-\lim_{t\to\infty}\frac{\lambda}{t}\log E[e^{-(\lambda u)\int_0^t r(\tau)\mathrm{d}\tau}]}{\lambda u} \\
\overset{(c)}{=}\;& \lambda \times \alpha_{K,\zeta,\beta}(\lambda u)
\end{aligned}
\tag{301}
$$

where (a) from (28), (b) using $\hat{r}(t) = \lambda r(t)$, which is obtained via (120), and (c) from (300). Then by $\alpha_{K,\lambda\zeta,\lambda\beta}(u) = \lambda \times \alpha_{K,\zeta,\beta}(\lambda u) \doteq \lambda\mu$, we have

$$
u = \alpha_{K,\lambda\zeta,\lambda\beta}^{-1}(\lambda\mu)
\tag{302}
$$

and

$$
\lambda u = \alpha_{K,\zeta,\beta}^{-1}(\mu).
\tag{303}
$$

Removing u in (302) and (303) results in

$$
\alpha_{K,\zeta,\beta}^{-1}(\mu) = \lambda \times \alpha_{K,\lambda\zeta,\lambda\beta}^{-1}(\lambda\mu)
\tag{304}
$$

Thus, we have

$$
\begin{aligned}
\theta_{K,\zeta,\beta}(\mu) \overset{(a)}{=}\;& \mu \times \alpha_{K,\zeta,\beta}^{-1}(\mu) \\
\overset{(b)}{=}\;& \mu \times \lambda \times \alpha_{K,\lambda\zeta,\lambda\beta}^{-1}(\lambda\mu) \\
\overset{(c)}{=}\;& \theta_{K,\lambda\zeta,\lambda\beta}(\lambda\mu)
\end{aligned}
\tag{305}
$$

where (a) from (30), (b) from (304), and (c) from (30). This completes the proof. ∎

A.13 Proof of Proposition 6.1

We prove it by contradiction. Suppose there exists a channel assignment $\{w_{k,n}(t) : k = 1, \cdots, K; n = 1, \cdots, N\}$ such that $\sum_{k=1}^{K} \sum_{n=1}^{N} w_{k,n}(t) < N$ and $\sum_{n=1}^{N} w_{k,n}(t) c_{k,n}(t) \geq \sum_{n=1}^{N} \mathbf{1}(k = k^*(n, t)) c_{k,n}(t), \forall k$ where $k^*(n, t)$ is the index of the user whose capacity $c_{k,n}(t)$ is the largest among K users, for channel n, and $\mathbf{1}(\cdot)$ is an indicator function such that $\mathbf{1}(k = a) = 1$ if $k = a$, and $\mathbf{1}(k = a) = 0$ if $k \neq a$. Since $\sum_{k=1}^{K} \sum_{n=1}^{N} w_{k,n}(t) < N$ and $\sum_{k=1}^{K} w_{k,n}(t) \leq 1, \forall n$ [see (140)], there must exist $\{\tilde{w}_{k,n}(t) : k = 1, \cdots, K; n = 1, \cdots, N\}$ such that $\sum_{k=1}^{K} \sum_{n=1}^{N} \tilde{w}_{k,n}(t) = N$, $\sum_{k=1}^{K} \tilde{w}_{k,n}(t) = 1, \forall n$ and $\tilde{w}_{k,n}(t) \geq w_{k,n}(t)$. Then, we have

$$\sum_{k=1}^{K} \sum_{n=1}^{N} \tilde{w}_{k,n}(t) c_{k,n}(t) \quad > \quad \sum_{k=1}^{K} \sum_{n=1}^{N} w_{k,n}(t) c_{k,n}(t) \tag{306}$$

$$\geq \quad \sum_{k=1}^{K} \sum_{n=1}^{N} \mathbf{1}(k = k^*(n, t)) c_{k,n}(t) \tag{307}$$

$$= \quad \sum_{n=1}^{N} c_{k^*(n,t),n}(t) \tag{308}$$

That is,

$$\sum_{k=1}^{K} \sum_{n=1}^{N} \tilde{w}_{k,n}(t) c_{k,n}(t) \quad > \quad \sum_{n=1}^{N} c_{k^*(n,t),n}(t) \tag{309}$$

On the other hand, since $\sum_{k=1}^{K} \tilde{w}_{k,n}(t) = 1, \forall n$, we have $\sum_{k=1}^{K} \tilde{w}_{k,n}(t) c_{k,n}(t) \leq c_{k^*(n,t),n}(t), \forall n$. Hence,

$$\sum_{n=1}^{N} \sum_{k=1}^{K} \tilde{w}_{k,n}(t) c_{k,n}(t) \quad \leq \quad \sum_{n=1}^{N} c_{k^*(n,t),n}(t) \tag{310}$$

(309) and (310) are contradictory. ∎

A.14 Proof of Proposition 6.2

By definition of $k^*(n,t)$, the capacities $c_{k^*(n,t),n}(t)$ are independent of $\{c_{k,n}(t), k \leq K\}$, and hence is independent of $\{w_{k,n}(t), k \leq K\}$. Thus, (146) becomes

$$
\begin{aligned}
C_{exp} &= \sum_{n=1}^{N} \left[\left(1 - \mathbf{E}\left[\sum_{k=1}^{K} w_{k,n}(t) \right] \right) \mathbf{E} c_{k^*(n,t),n}(t) \right] \\
&\stackrel{(a)}{=} \mathbf{E}[c_{k^*(n,t),n}(t)] \times \left(N - \mathbf{E} \sum_{n=1}^{N} \sum_{k=1}^{K} w_{k,n}(t) \right) \\
&= \mathbf{E}[c_{k^*(n,t),n}(t)] \times (N - N \times \eta(K,N))
\end{aligned}
$$

where (a) is due to the fact that $c_{k,n}(t)$ $(k = K+1, \cdots, K+K_B)$ are i.i.d. and strict-sense stationary, and hence $c_{k^*(n,t),n}(t)$ are i.i.d and strict-sense stationary. Therefore, minimizing the expected channel usage $\eta(K,N)$ is equivalent to maximizing the available expected capacity C_{exp}. ∎

A.15 Proof of Proposition 6.3

It is clear that the minimum value of the objective (138) under the constraint of (139) and (141) is a lower bound on that of (138) under the constraints of (139) through (141). The solution for (138), (139) and (141), is simply that each user only chooses its best channel to transmit (even though the total usage of a channel by all users could be more than 1), i.e.,

$$
w_{k,n}(t) = \frac{\sum_{m=1}^{N} \xi_{k,m} c_{k,m}(t)}{c_{k,n}(t)} \times \mathbf{1}(n = \bar{n}(k,t)), \qquad \forall k, \forall n \tag{311}
$$

where $\bar{n}(k,t)$ is the index of the channel whose capacity $c_{k,n}(t)$ is the largest among N channels for user k. So we get $\eta(K,N)$ for the scheduler specified by (138) through (141) as below,

$$
\begin{aligned}
\eta(K,N) &\stackrel{(a)}{=} \frac{\mathbf{E}[\sum_{k=1}^{K} \sum_{n=1}^{N} w_{k,n}(t)]}{N} \\
&\stackrel{(b)}{\geq} \frac{\mathbf{E}[\sum_{k=1}^{K} (\frac{\sum_{n=1}^{N} \xi_{k,n} c_{k,n}(t)}{c_{k,\bar{n}(k,t)}(t)})]}{N} \\
&\stackrel{(c)}{=} \frac{(\sum_{k=1}^{K} N\xi_{k,n})\mathbf{E}[\frac{\sum_{n=1}^{N} c_{k,n}/N}{c_{max}}]}{N} \\
&\stackrel{(d)}{=} \mathbf{E}[\frac{c_{mean}}{c_{max}}]
\end{aligned}
$$

where (a) due to the fact that $c_{k,n}(t)$ are stationary, thereby $w_{k,n}(t)$ being stationary, (b) since the assignment in (311) gives a lower bound, (c) since $c_{k,n}(t)$ are i.i.d. and stationary, and (d) due to (147). This completes the proof. ∎

A.16   Proof of Proposition 6.4

We first present a lemma and then prove Proposition 6.4.

Let $\gamma = P_0/\sigma^2$. Denote g_1 and g_2 channel power gains of two fading channels, respectively. Lemma A.1 says that for fixed channel gain ratio $g_1/g_2 > 1$, the corresponding capacity ratio $\log(1+\gamma \times g_2)/\log(1+\gamma \times g_1)$ monotonically increases from g_2/g_1 to 1, as average SNR γ increases from 0 to ∞.

Lemma A.1 *If $g_1 > g_2 > 0$, then $\log(1+\gamma \times g_2)/\log(1+\gamma \times g_1)$ monotonically increases from g_2/g_1 to 1, as γ increases from 0 to ∞.*

Proof: We prove it by considering three cases: 1) $0 < \gamma < \infty$, 2) $\gamma = 0$, and 3) $\gamma \to \infty$.

Case 1: $0 < \gamma < \infty$

Define $f(\gamma) = \log(1 + \gamma g_2)/\log(1 + \gamma g_1)$. To prove the lemma for Case 1, we only need to show $f'(\gamma) > 0$ for $\gamma > 0$. Taking the derivative results in

$$f'(\gamma) = \frac{\frac{g_2}{1+\gamma g_2}\log(1+\gamma g_1) - \frac{g_1}{1+\gamma g_1}\log(1+\gamma g_2)}{\log^2(1+\gamma g_1)} \quad (312)$$

Since $\log^2(1+\gamma g_1) > 0$ for $\gamma > 0$, we only need to show

$$\frac{g_2}{1+\gamma g_2}\log(1+\gamma g_1) > \frac{g_1}{1+\gamma g_1}\log(1+\gamma g_2) \quad (313)$$

or equivalently, that,

$$\frac{\frac{g_2}{(1+\gamma g_2)\log(1+\gamma g_2)}}{\frac{g_1}{(1+\gamma g_1)\log(1+\gamma g_1)}} > 1 \quad (314)$$

Define $h(x) = \frac{x}{(1+\gamma x)\log(1+\gamma x)}$. If $h'(x) < 0$ for $x > 0$, then $g_1 > g_2 > 0$ implies $0 < h(g_1) < h(g_2)$, *i.e.*, $h(g_2)/h(g_1) > 1$, which is the inequality in (314). So we only need to show $h'(x) < 0$ for $x > 0$. Taking the derivative, we have

$$h'(x) = \frac{\frac{1+\gamma x - \gamma x}{(1+\gamma x)^2}\log(1+\gamma x) - \frac{\gamma}{1+\gamma x}\frac{x}{1+\gamma x}}{\log^2(1+\gamma x)} \quad (315)$$

$$= \frac{\frac{1}{(1+\gamma x)^2}(\log(1+\gamma x) - \gamma x)}{\log^2(1+\gamma x)} \quad (316)$$

258

For $\gamma > 0$ and $x > 0$, we have $\log(1 + \gamma x) - \gamma x < 0$, which implies $h'(x) < 0$.

Case 2: $\gamma = 0$

$$\lim_{\gamma \to 0} f(\gamma) \overset{(a)}{=} \lim_{\gamma \to 0} \frac{\frac{g_2}{1+\gamma g_2}}{\frac{g_1}{1+\gamma g_1}} = \frac{g_2}{g_1} \tag{317}$$

where (a) is from L'Hospital's rule.

Case 3: $\gamma \to \infty$

$$\lim_{\gamma \to \infty} f(\gamma) \overset{(a)}{=} \lim_{\gamma \to \infty} \frac{\frac{g_2}{1+\gamma g_2}}{\frac{g_1}{1+\gamma g_1}} \tag{318}$$

$$= \lim_{\gamma \to \infty} \frac{g_2(1 + \gamma g_1)}{g_1(1 + \gamma g_2)} \tag{319}$$

$$\overset{(b)}{=} \frac{g_2}{g_1} \times \frac{g_1}{g_2} \tag{320}$$

$$= 1 \tag{321}$$

where (a) and (b) are from L'Hospital's rule.

Combining Cases 1 to 3, we complete the proof. ∎

Next, we prove Proposition 6.4.

Proof of Proposition 6.4: From the definition of c_{max}, we have

$$\begin{aligned} c_{max} &= \max_{n \in \{1,2,\cdots,N\}} c_{1,n} \\ &= \max_{n \in \{1,2,\cdots,N\}} \log(1 + \gamma \times g_{1,n}) \\ &= \log(1 + \gamma \times g_{max}) \end{aligned} \tag{322}$$

where $g_{max} = \max_{n \in \{1,2,\cdots,N\}} g_{1,n}$. Also, from the definition of c_{mean}, we get

$$\begin{aligned} c_{mean} &= \frac{1}{N} \sum_{n=1}^{N} c_{1,n} \\ &= \frac{1}{N} \sum_{n=1}^{N} \log(1 + \gamma \times g_{1,n}) \\ &= \log(1 + \gamma \times g_{mean}) \end{aligned} \tag{323}$$

where

$$g_{mean} = \frac{1}{\gamma} (\prod_{n=1}^{N} (1 + \gamma \times g_{1,n})^{1/N} - 1) \tag{324}$$

259

It is obvious that $g_{max} > g_{mean} > 0$. So from Lemma A.1, we have $\log(1 + \gamma \times g_{mean})/\log(1 + \gamma \times g_{max})$, *i.e.*, c_{mean}/c_{max}, monotonically increases from g_{mean}/g_{max} to 1, as γ increases from 0 to ∞. Hence, $\mathbf{E}[c_{mean}/c_{max}]$ monotonically increases from $\mathbf{E}[g_{mean}/g_{max}]$ to 1, as γ increases from 0 to ∞. ■

A.17 Facts

Fact A.1 *Suppose that events $\mathcal{A}_i$ form a partition of event $\mathcal{A}$ and events $\mathcal{B}_{ij}$ form a partition of $\mathcal{A}_i$. Denote p_i the probability of event $\mathcal{A}_i$ and q_j the conditional probability of an event $\mathcal{B}_{ij}$ assuming $\mathcal{A}_i$. Define a random variable $\mathbf{x}$ on $\mathcal{A}$ and a random variable $\mathbf{y}_i$ on $\mathcal{A}_i$. Then we have*

$$E\{\mathbf{x}\} = \sum_i (p_i \cdot E\{\mathbf{y}_i\}). \tag{325}$$

A.18 Proof of Proposition 8.1

From Eq. (208), one can easily obtain Eq. (211). Note that due to the random nature of the channel characteristics, $\hat{f}_{sj}^N$ at the receiver is a random variable while f_{sj}^N is not a random variable.

To see that Eq. (212) holds, we consider the following two cases based on the position of F_i^n.

- *Case 1:* Suppose that F_i^n only has $F_{\bar{i}}^n$ or $F_{\tilde{i}}^n$ (*e.g.*, F_i^n is located at the top/bottom of the frame), *i.e.*, $F_i^n \in \mathcal{G}^n$. According to the packet loss behavior, there are three cases for the random variable $\hat{f}_{ij}^n$ as follows.

  - *Subcase 1.1:* Suppose that $\bar{F}_i^n$ is received correctly. Then we have $\hat{f}_{ij}^n = \tilde{f}_{ij}^n$, and the probability of this event is $1 - p_d$.

  - *Subcase 1.2:* For simplicity, we only show the situation where F_i^n is located at the top of the frame. The situation where F_i^n is located at the bottom of the frame is the same.
    Suppose that $\bar{F}_i^n$ is corrupted and the MB below (*i.e.*, $\bar{F}_{\tilde{i}}^n$) has been received correctly. Then due to our error concealment scheme, the corrupted pixel j in F_i^n is concealed by the pixel in frame $n-1$, which is pointed by the motion vector of $\bar{F}_{\tilde{i}}^n$. Since this pixel is denoted by pixel l in macroblock m of frame $n - 1$ (see Table 3), then we have $\hat{f}_{ij}^n = \hat{f}_{ml}^{n-1}$, and the probability of this event is $p_d \cdot (1 - p_d)$. Since $\hat{f}_{ml}^{n-1}$ is also a random variable, from Fact A.1 (see Appendix A.17), the

expectation of $\hat{f}_{ml}^{n-1}$ should be chosen in computing the expectation of $\hat{f}_{ij}^{n}$. So we have $\hat{f}_{ij}^{n} = E\{\hat{f}_{ml}^{n-1}\}$.

– *Subcase 1.3:* For simplicity, we only show the situation where F_{i}^{n} is located at the top of the frame. The situation where F_{i}^{n} is located at the bottom of the frame is the same.

Suppose that $\bar{F}_{i}^{n}$ and $\bar{F}_{\bar{i}}^{n}$ are all corrupted. Since $\bar{F}_{\bar{i}}^{n}$ is corrupted, the estimated motion vector of the corrupted F_{i}^{n} is set to zero. Then the corrupted pixel j in F_{i}^{n} is concealed by pixel j in F_{i}^{n-1}. Thus, we have $\hat{f}_{ij}^{n} = \hat{f}_{ij}^{n-1}$, and the probability of this event is p_{d}^{2}. Since $\hat{f}_{ij}^{n-1}$ is also a random variable, from Fact A.1, the expectation of $\hat{f}_{ij}^{n-1}$ should be chosen in computing the expectation of $\hat{f}_{ij}^{n}$. So we have $\hat{f}_{ij}^{n} = E\{\hat{f}_{ij}^{n-1}\}$.

Based on the analysis of Subcases 1.1 to 1.3, the expectation of $\hat{f}_{ij}^{n}$ can be given by

$$E\{\hat{f}_{ij}^{n}\} = (1 - p_{d}) \cdot \tilde{f}_{ij}^{n} + p_{d} \cdot (1 - p_{d}) \cdot E\{\hat{f}_{ml}^{n-1}\} + p_{d}^{2} \cdot E\{\hat{f}_{ij}^{n-1}\}. \quad (326)$$

- *Case 2:* Suppose that F_{i}^{n} has both $F_{\bar{i}}^{n}$ and $F_{\bar{i}}^{n}$ (*e.g.*, F_{i}^{n} is not located at the top/bottom of the frame), *i.e.*, $F_{i}^{n} \notin \mathcal{G}^{n}$.

According to the packet loss behavior, there are three cases for the random variable $\hat{f}_{ij}^{n}$ as follows.

– *Subcase 2.1:* Suppose that $\bar{F}_{i}^{n}$ is received correctly. Then we have $\hat{f}_{ij}^{n} = \tilde{f}_{ij}^{n}$, and the probability of this event is $1 - p_{d}$.

– *Subcase 2.2:* Suppose that $\bar{F}_{i}^{n}$ is corrupted and the MB above and/or below (*i.e.*, $\bar{F}_{\bar{i}}^{n}$ and/or $\bar{F}_{\bar{i}}^{n}$) have/has been received correctly. Then due to our error concealment scheme, the corrupted pixel j in F_{i}^{n} is concealed by the pixel in frame $n - 1$, which is pointed by the motion vector of $\bar{F}_{\bar{i}}^{n}$ or $\bar{F}_{\bar{i}}^{n}$. Since this pixel is denoted by pixel l in macroblock m of frame $n - 1$, then we have $\hat{f}_{ij}^{n} = \hat{f}_{ml}^{n-1}$, and the probability of this event is $p_{d} \cdot (1 - p_{d}^{2})$. Since $\hat{f}_{ml}^{n-1}$ is also a random variable, from Fact A.1, the expectation of $\hat{f}_{ml}^{n-1}$ should be chosen in computing the expectation of $\hat{f}_{ij}^{n}$. So we have $\hat{f}_{ij}^{n} = E\{\hat{f}_{ml}^{n-1}\}$.

– *Subcase 2.3:* Suppose that $\bar{F}_{i}^{n}$, $\bar{F}_{\bar{i}}^{n}$, and $\bar{F}_{\bar{i}}^{n}$ are all corrupted. Since the MBs above and below F_{i}^{n} are corrupted, the estimated motion

261

vector of the corrupted F_i^n is set to zero. Then the corrupted pixel j in F_i^n is concealed by pixel j in F_i^{n-1}. Thus, we have $\hat{f}_{ij}^n = \hat{f}_{ij}^{n-1}$, and the probability of this event is p_d^3. Since $\hat{f}_{ij}^{n-1}$ is also a random variable, from Fact A.1, the expectation of $\hat{f}_{ij}^{n-1}$ should be chosen in computing the expectation of $\hat{f}_{ij}^n$. So we have $\hat{f}_{ij}^n = E\{\hat{f}_{ij}^{n-1}\}$.

Based on the analysis of Subcases 2.1 to 2.3, the expectation of $\hat{f}_{ij}^n$ can be given by

$$E[\hat{f}_{ij}^n] = (1 - p_d) \cdot \tilde{f}_{ij}^n + p_d \cdot (1 - p_d^2) \cdot E\{\hat{f}_{ml}^{n-1}\} + p_d^3 \cdot E\{\hat{f}_{ij}^{n-1}\} \quad (327)$$

Combining Cases 1 and 2 above, we complete the proof of Eq. (212). The proof for Eq. (213) is similar to that for Eq. (212). ∎

A.19 Proof of Proposition 8.2

Similar to the derivation of Eq. (211), we have Eq. (214). To see that Eqs. (215) and (216) hold, we can also use the same way as that to derive Eqs. (212) and (213). The only difference is the case where $\bar{F}_i^n$ is received correctly. In this case, under the inter code, we have $\hat{f}_{ij}^n = \tilde{e}_{ij}^n + \hat{f}_{uv}^{n-1}$, which is Eq. (210). ∎

BIBLIOGRAPHY

[1] M. Ahmed, H. Yanikomeroglu, and S. Mahmoud, "Call admission control in wireless communications: a comprehensive survey," to be submitted to *IEEE Wireless Communications Magazine.*

[2] A. Albanese, J. Blömer, J. Edmonds, M. Luby, and M. Sudan, "Priority encoding transmission," *IEEE Trans. on Information Theory,* vol. 42, no. 6, pp. 1737–1744, Nov. 1996.

[3] A. Alwan, R. Bagrodia, N. Bambos, M. Gerla, L. Kleinrock, J. Short, and J. Villasenor, "Adaptive mobile multimedia networks," *IEEE Personal Communications Magazine,* vol. 3, no. 2, pp. 34–51, April 1996.

[4] M. Andrews, K. Kumaran, K. Ramanan, A. Stolyar, P. Whiting, and R. Vijayakumar, "Providing quality of service over a shared wireless link," *IEEE Communications Magazine,* vol. 39, no. 2, pp. 150–154, Feb. 2001.

[5] ATM Forum Technical Committee, "Traffic management specification (version 4.0)," ATM Forum, Feb. 1996.

[6] A. Balachandran, A. T. Campbell, M. E. Kounavis, "Active filters: delivering scalable media to mobile devices," in *Proc. Seventh International Workshop on Network and Operating System Support for Digital Audio and Video (NOSSDAV'97),* St Louis, MO, USA, May 1997.

[7] N. Bambos and S. Kandukuri, "Power-controlled multiple access schemes for next-generation wireless packet networks," *IEEE Wireless Communications Magazine,* vol. 9, no. 3, pp. 58–64, June 2002.

[8] D. Bertsekas, *Dynamic Programming and Optimal Control, Vol. 1, 2,* Athena Scientific, 1995.

[9] I. Bettesh and S. Shamai, "A low delay algorithm for the multiple access channel with Rayleigh fading," in *Proc. IEEE Personal, Indoor and Mobile Radio Communications (PIMRC'98),* 1998.

[10] I. Bettesh and S. Shamai, "Optimal power and rate control for fading channels," in *Proc. IEEE Vehicular Technology Conference,* Spring 2001.

[11] C. Bettstetter, "Smooth is better than sharp: a random mobility model for simulation of wireless networks," in *Proc. 4th ACM International Workshop on Modeling, Analysis, and Simulation of Wireless and Mobile Systems (MSWiM),* Rome, Italy, July 2001.

[12] G. Bianchi, A. T. Campbell, and R. Liao, "On utility-fair adaptive services in wireless networks," in *Proc. 6th International Workshop on Quality of Service (IWQOS'98),* Napa Valley, CA, May 1998.

[13] E. Biglieri, J. Proakis, and S. Shamai, "Fading channel: information theoretic and communication aspects," *IEEE Trans. Information Theory,* vol. 44, pp. 2619–2692, Oct. 1998.

[14] S. Blake, D. Black, M. Carlson, E. Davies, Z. Wang, and W. Weiss, "An architecture for differentiated services," *RFC 2475,* Internet Engineering Task Force, Dec. 1998.

[15] J-C. Bolot, T. Turletti, and I. Wakeman, "Scalable feedback control for multicast video distribution in the Internet," in *Proc. ACM SIGCOMM'94,* pp. 58–67, London, UK, Sept. 1994.

[16] J-C. Bolot and T. Turletti, "Adaptive error control for packet video in the Internet," in *Proc. IEEE Int. Conf. on Image Processing (ICIP'96),* pp. 25–28, Lausanne, Switzerland, Sept. 1996.

[17] J-C. Bolot and T. Turletti, "Experience with control mechanisms for packet video in the Internet," *ACM Computer Communication Review,* vol. 28, no. 1, Jan. 1998.

[18] D. Botvich and N. Duffield, "Large deviations, the shape of the loss curve, and economies of scale in large multiplexers," *Queueing Systems,* vol. 20, pp. 293–320, 1995.

[19] R. Braden, D. Clark, and S. Shenker, "Integrated services in the Internet architecture: An overview," *RFC 1633,* Internet Engineering Task Force, July 1994.

[20] R. Braden (Ed.), L. Zhang, S. Berson, S. Herzog, and S. Jamin, "Resource ReSerVation Protocol (RSVP) version 1, functional specification," *RFC 2205,* Internet Engineering Task Force, Sept. 1997.

[21] B. Braden, D. Black, J. Crowcroft, B. Davie, S. Deering, D. Estrin, S. Floyd, V. Jacobson, G. Minshall, C. Partridge, L. Peterson, K. Ramakrishnan, S. Shenker, J. Wroclawski, and L. Zhang, "Recommendations on queue management and congestion avoidance in the Internet," *RFC 2309,* Internet Engineering Task Force, April 1998.

[22] C.-S. Chang, "Stability, queue length and delay of deterministic and stochastic queueing networks," *IEEE Transactions on Automatic Control,* vol. 39, no. 5, pp. 913–931, May 1994.

[23] C.-S. Chang and J. A. Thomas, "Effective bandwidth in high-speed digital networks," *IEEE Journal on Selected Areas in Communications,* vol. 13, no. 6, pp. 1091–1100, Aug. 1995.

[24] C.-S. Chang, "Performance guarantees in communication networks," Springer, 2000.

[25] S. Y. Cheung, M. Ammar, and X. Li, "On the use of destination set grouping to improve fairness in multicast video distribution," in *Proc. IEEE INFO-COM'96,* pp. 553–560, San Francisco, CA, March 1996.

[26] T. Chiang and Y.-Q. Zhang, "A new rate control scheme using quadratic rate distortion model," *IEEE Trans. on Circuits and Systems for Video Technology,* vol. 7, no. 1, pp. 246–250, Feb. 1997.

[27] P. A. Chou, "Joint source/channel coding: a position paper," in *Proc. NSF Workshop on Source-Channel Coding,* San Diego, CA, USA, Oct. 1999.

[28] G. L. Choudhury, D. M. Lucantoni, W. Whitt, "Squeezing the most out of ATM," *IEEE Transactions on Communications,* vol. 44, no. 2, pp. 203–217, Feb. 1996.

[29] D. Clark, S. Shenker, and L. Zhang, "Supporting real-time applications in an integrated services packet network: architecture and mechanisms," in *Proc. ACM SIGCOMM'92,* Baltimore, MD, Aug. 1992.

[30] B. E. Collins and R. L. Cruz, "Transmission policies for time-varying channels with average delay constraints," in *Proc. Allerton Conference on Communication, Control, and Computing,* Monticello IL, Sept. 1999.

[31] G. Cote and F. Kossentini, "Optimal intra coding of blocks for robust video communication over the Internet," to appear in *EUROSIP Image Communication Special Issue on Real-time Video over the Internet.*

[32] G. Cote, S. Shirani, and F. Kossentini, "Optimal mode selection and synchronization for robust video communications over error-prone networks," *IEEE Journal on Selected Areas in Communications,* vol. 18, no. 6, pp. 952–965, June 2000.

[33] T. Cover and J. Thomas, *Elements of information theory,* John Wiley & Sons, New York, 1991.

[34] R. L. Cruz, "A calculus for network delay, Part I: network elements in isolation," *IEEE Trans. on Information Theory,* vol. 37, no. 1, pp. 114–131, Jan. 1991.

[35] G. Davis and J. Danskin, "Joint source and channel coding for Internet image transmission," in *Proc. SPIE Conference on Wavelet Applications of Digital Image Processing XIX,* Denver, Aug. 1996.

[36] Kalyanmoy Deb, "Multi-objective optimization using evolutionary algorithms," John Wiley & Sons, 2001.

[37] B. J. Dempsey, J. Liebeherr, and A. C. Weaver, "On retransmission-based error control for continuous media traffic in packet-switching networks," *Computer Networks and ISDN Systems,* vol. 28, no. 5, pp. 719–736, March 1996.

[38] G. de Veciana and J. Walrand, "Effective bandwidths: call admission, traffic policing and filtering for ATM networks," *Queuing Systems,* vol. 20, pp. 37–59, 1995.

[39] W. Ding and B. Liu, "Rate control of MPEG video coding and recording by rate-quantization modeling," *IEEE Trans. on Circuits and Systems for Video Technology,* vol. 6, pp. 12–20, Feb. 1996.

[40] W. Ding, "Joint encoder and channel rate control of VBR video over ATM networks," *IEEE Trans. on Circuits and Systems for Video Technology,* vol. 7, pp. 266–278, April 1997.

[41] A. Dornan, *The essential guide to wireless communications applications,* Prentice Hall, Upper Saddle River, NJ, 2002.

[42] A. E. Eckberg, "Approximations for bursty and smoothed arrival delays based on generalized peakedness," in *Proc. 11th International Teletraffic Congress,* Kyoto, Japan, Sept. 1985.

[43] M. Elaoud and P. Ramanathan, "Adaptive allocation of CDMA resources for network level QoS assurances," in *Proc. ACM Mobicom'00,* Aug. 2000.

[44] A. Eleftheriadis and D. Anastassiou, "Meeting arbitrary QoS constraints using dynamic rate shaping of coded digital video," in *Proc. IEEE Int. Workshop on Network and Operating System Support for Digital Audio and Video (NOSSDAV'95),* pp. 95–106, April 1995.

[45] J. S. Evans and D. Everitt, "Effective bandwidth-based admission control for multiservice CDMA cellular networks," *IEEE Trans. on Vehicular Technology,* vol. 48, no. 1, pp. 36–46, Jan. 1999.

[46] H. Fattah and C. Leung, "An overview of scheduling algorithms in wireless multimedia networks," *IEEE Wireless Communications Magazine,* vol. 9, no. 5, pp. 76–83, Oct. 2002.

[47] S. Floyd, and K. Fall, "Promoting the use of end-to-end congestion control in the Internet," *IEEE/ACM Trans. on Networking,* vol. 7, no. 4, pp. 458–472, Aug. 1999.

[48] P. Frenger, P. Orten, T. Ottosson, and A. Svensson, "Multi-rate convolutional codes," *Tech. Report No. 21,* Dept. of Signals and Systems, Chalmers University of Technology, page 131, April 1998.

[49] L. Georgiadis, R. Guerin, V. Peris, and R. Rajan, "Efficient support of delay and rate guarantees in an Internet," in *Proc. ACM SIGCOMM'96,* Aug. 1996.

[50] M. Ghanbari, "Cell-loss concealment in ATM video codes," *IEEE Trans. on Circuits and Systems for Video Technology,* vol. 3, pp. 238–247, June 1993.

[51] M. Grossglauser, S. Keshav, and D. Tse, "RCBR: a simple and efficient service for multiple time-scale traffic," *IEEE/ACM Trans. on Networking,* vol. 5, no. 6, pp. 741–755, Dec. 1997.

[52] M. Grossglauser and D. Tse, "Mobility increases the capacity of wireless adhoc networks," in *Proc. IEEE INFOCOM'01,* April 2001.

[53] M. Gudmundson, "Correlation model for shadow fading in mobile radio systems," *IEE Electronics Letters,* vol. 27, no. 23, pp. 2145–2146, Nov. 1991.

[54] R. Guerin and V. Peris, "Quality-of-service in packet networks: basic mechanisms and directions," *Computer Networks and ISDN,* vol. 31, no. 3, pp. 169–179, Feb. 1999.

[55] J. Hagenauer, "Rate-compatible punctured convolutional codes (RCPC codes) and their applications," *IEEE Trans. Commun.,* vol. 36, pp. 389–400, April 1988.

[56] J. Hagenauer, "Source-controlled channel decoding," *IEEE Trans. on Communications,* vol. 43, no. 9, pp. 2449–2457, Sept. 1995.

[57] S. Hanly and D. Tse, "Multi-access fading channels: part II: delay-limited capacities," *IEEE Trans. on Information Theory,* vol. 44, no. 7, pp. 2816–2831, Nov. 1998.

[58] J. Heinanen, F. Baker, W. Weiss and J. Wroclawski, "Assured forwarding PHB group," *RFC 2597,* Internet Engineering Task Force, June 1999.

[59] M. Hemy, U. Hengartner, P. Steenkiste, and T. Gross, "MPEG system streams in best-effort networks," in *Proc. IEEE Packet Video'99,* New York, April 26–27, 1999.

[60] H. Holma and A. Toskala, *WCDMA for UMTS: Radio Access for Third Generation Mobile Communications,* Wiley, 2000.

[61] L. M. C. Hoo, "Multiuser transmit optimization for multicarrier modulation system," *Ph. D. Dissertation,* Department of Electrical Engineering, Stanford University, CA, USA, Dec. 2000.

[62] C. Y. Hsu, A. Ortega, and A. R. Reibman, "Joint selection of source and channel rate for VBR transmission under ATM policing constraints," *IEEE Journal on Selected Areas in Communications,* vol. 15, pp. 1016–1028, Aug. 1997.

[63] A. Iera, A. Molinaro, and S. Marano, "Wireless broadband applications: the teleservice model and adaptive QoS provisioning," *IEEE Communications Magazine,* vol. 37, no. 10, pp. 71–75, Oct. 1999.

[64] ISO/IEC JTC 1/SC 29/WG 11, "Information technology – coding of audio-visual objects, part 1: systems, part 2: visual, part 3: audio," *FCD 14496,* Dec. 1998.

[65] B. Jabbari, "Teletraffic aspects of evolving and next-generation wireless communication networks," *IEEE Personal Communications Magazine,* pp. 4–9, Dec. 1996.

[66] S. Jacobs and A. Eleftheriadis, "Streaming video using TCP flow control and dynamic rate shaping," *Journal of Visual Communication and Image Representation,* vol. 9, no. 3, pp. 211–222, Sept. 1998.

[67] V. Jacobson, "Congestion avoidance and control," in *Proc. ACM SIGCOMM'88,* pp. 314–329, Aug. 1988.

[68] V. Jacobson, K. Nichols, and K. Poduri, "An expedited forwarding PHB," *RFC 2598,* Internet Engineering Task Force, June 1999.

[69] S. Jamin, P. B. Danzig, S. Shenker and L. Zhang, "A measurement-based admission control algorithm for integrated services packet networks," *IEEE/ACM Trans. on Networking,* vol. 5, no. 1, pp. 56–70, Feb. 1997.

[70] S. Kallel and D. Haccoun, "Generalized type-II hybrid ARQ scheme using punctured convolutional coding," *IEEE Trans. Commun.,* vol. 38, pp. 1938–1946, Nov. 1990.

[71] Y. Y. Kim and S.-Q. Li, "Capturing important statistics of a fading/shadowing channel for network performance analysis," *IEEE Journal on Selected Areas in Communications,* vol. 17, no. 5, pp. 888–901, May 1999.

[72] R. Knopp and P. A. Humblet, "Information capacity and power control in single-cell multiuser communications," in *Proc. IEEE International Conference on Communications (ICC'95),* Seattle, USA, June 1995.

[73] J. Knutsson, P. Butovitsch, M. Persson, R. D. Yates, "Downlink admission control strategies for CDMA systems in a Manhattan environment," *Proc. IEEE 48th Vehicular Technology Conference (VTC98),* May 1998.

[74] J. Kuri and P. Mermelstein, "Call admission on the uplink of a CDMA system based on total received power," *Proc. IEEE International Conference on Communications (ICC'99),* June 1999.

268

[75] O. Lataoui, T. Rachidi, L. G. Samuel, S. Gruhl, and R.-H., Yan, "A QoS management architecture for packet switched 3rd generation mobile systems," in *Proc. Networld+Interop 2000 Engineers Conference,* Las Vegas, Nevada, USA, May 10–11, 2000.

[76] J.-Y. Le Boudec and P. Thiran, "Network calculus: a theory of deterministic queueing systems for the Internet," Springer, 2001.

[77] J. Lee and B.W. Dickenson, "Rate-distortion optimized frame type selection for MPEG encoding," *IEEE Trans. on Circuits and Systems for Video Technology,* vol. 7, pp. 501–510, June 1997.

[78] K. Lee, "Adaptive network support for mobile multimedia," in *Proc. ACM Mobicom'95,* Berkeley, CA, USA, Nov. 13–15, 1995.

[79] X. Li, S. Paul, P. Pancha, and M. H. Ammar, "Layered video multicast with retransmissions (LVMR): evaluation of error recovery schemes," in *Proc. IEEE Int. Workshop on Network and Operating System Support for Digital Audio and Video (NOSSDAV'97),* pp. 161–172, May 1997.

[80] X. Li, S. Paul, and M. H. Ammar, "Layered video multicast with retransmissions (LVMR): evaluation of hierarchical rate control," in *Proc. IEEE INFOCOM'98,* vol. 3, pp. 1062–1072, March 1998.

[81] S. Lin and D. Costello, "Error control coding: fundamentals and applications," Englewood Cliffs, NJ: Prentice-Hall, 1983.

[82] H. Liu and M. El Zarki, "Performance of H.263 video transmission over wireless channels using hybrid ARQ," *IEEE J. on Selected Areas in Communications,* vol. 15, no. 9, pp. 1775–1786, Dec. 1997.

[83] X. Liu, E. K. P. Chong, and N. B. Shroff, "Opportunistic transmission scheduling with resource-sharing constraints in wireless networks," *IEEE Journal on Selected Areas in Communications,* vol. 19, no. 10, pp. 2053–2064, Oct. 2001.

[84] Z. Liu and M. El Zarki, "SIR-based call admission control for DS-CDMA cellular systems," *IEEE Journal on Selected Areas in Communications,* vol. 12, no. 4, pp. 638–644, May 1994.

[85] R. M. Loynes, "The stability of a queue with non-independent inter-arrivals and service times," *Proc. Camb. Phil. Soc.,* vol. 58, pp. 497–520, 1962.

[86] S. Lu, K.-W. Lee, and V. Bharghavan, "Adaptive service in mobile computing environments," in *Proc. 5th International Workshop on Quality of Service (IWQOS'97),* Columbia University, New York, May 21–23, 1997.

[87] S. Lu, V. Bharghavan, and R. Srikant, "Fair scheduling in wireless packet networks," *IEEE/ACM Trans. on Networking,* vol. 7, no. 4, pp. 473–489, Aug. 1999.

[88] B. L. Mark and G. Ramamurthy, "Real-time estimation and dynamic rene-gotiation of UPC parameters for arbitrary traffic sources in ATM networks," *IEEE/ACM Trans. on Networking,* vol. 6, no. 6, pp. 811–827, Dec. 1998.

[89] F. C. Martins, W. Ding, and E. Feig, "Joint control of spatial quantization and temporal sampling for very low bit-rate video," in *Proc. IEEE Int. Conference on Acoustics, Speech, and Signal Processing (ICASSP'96),* vol. 4, pp. 2072–2075, May 1996.

[90] N. Maxemchuk, K. Padmanabhan, and S. Lo, "A cooperative packet recovery protocol for multicast video," in *Proc. IEEE Int. Conference on Network Protocols (ICNP'97),* pp. 259–266, Oct. 1997.

[91] S. McCanne, V. Jacobson, and M. Vetterli, "Receiver-driven layered multi-cast," in *Proc. ACM SIGCOMM'96,* pp. 117–130, Aug. 1996.

[92] J. Mogul and S. Deering, "Path MTU discovery," *RFC 1191,* Internet Engi-neering Task Force, RFC 1191, Nov. 1990.

[93] M. Naghshineh and M. Willebeek-LeMair, "End-to-end QoS provisioning in multimedia wireless/mobile networks using an adaptive framework," *IEEE Communications Magazine,* vol. 35, no. 11, pp. 72–81, Nov. 1997.

[94] T. S. E. Ng, I. Stoica, and H. Zhang, "Packet fair queueing algorithms for wire-less networks with location-dependent errors," in *Proc. IEEE INFOCOM'98,* pp. 1103–1111, San Francisco, CA, USA, March 1998.

[95] K. Nichols, V. Jacobson, and L. Zhang, "A two-bit differentiated services architecture for the Internet," *RFC 2638,* Internet Engineering Task Force, July 1999.

[96] K. Nichols, S. Blake, F. Baker and D. Black, "Definition of the differentiated services field (DS field) in the IPv4 and IPv6 headers," *RFC 2474,* Internet Engineering Task Force, Dec. 1998.

[97] J. Nocedal and S. J. Wright, *Numerical optimization,* Springer, 1999.

[98] A. Ortega and K. Ramchandran, "Rate-distortion methods for image and video compression," *IEEE Signal Processing Magazine,* pp. 74–90, Nov. 1998.

[99] A. K. Parekh and R. G. Gallager, "A generalized processor sharing ap-proach to flow control in integrated services networks: the single node case," *IEEE/ACM Trans. on Networking,* vol. 1, no. 3, pp. 344–357, June 1993.

[100] M. Podolsky, M. Vetterli, and S. McCanne, "Limited retransmission of real-time layered multimedia," in *Proc. IEEE Workshop on Multimedia Signal Processing,* pp. 591–596, Dec. 1998.

[101] M. Podolsky, K. Yano, and S. McCanne, "A RTCP-based retransmission protocol for unicast RTP streaming multimedia," *Internet draft,* Internet Engineering Task Force, Oct. 1999, work in progress.

[102] B. Prabhakar, E. Uysal-Biyikoglu, and A. El Gamal, "Energy-efficient transmission over a wireless link via lazy packet scheduling," in *Proc. IEEE INFOCOM'01,* April 2001.

[103] J. Proakis, "Digital communications," 4th ed., McGraw-Hill, Aug. 2000.

[104] R. J. Punnoose, P. V. Nikitin, and D. D. Stancil, "Efficient simulation of Ricean fading within a packet simulator," in *Proc. IEEE Vehicular Technology Conference (VTC'2000),* Boston, MA, USA, Sept. 2000.

[105] D. Rajan, A. Sabharwal, and B. Aazhang, "Delay and rate constrained transmission policies over wireless channels," in *Proc. IEEE GLOBECOM'01,* Nov. 2001.

[106] P. Ramanathan and P. Agrawal, "Adapting packet fair queueing algorithms to wireless networks," in *Proc. ACM MOBICOM'98,* Oct. 1998.

[107] K. Ramchandran, A. Ortega and M. Vetterli, "Bit allocation for dependent quantization with applications to multiresolution and MPEG video coders," *IEEE Trans. on Image Processing,* vol. 37, pp. 533–545, Aug. 1994.

[108] K. R. Rao, "MPEG-4 — the emerging multimedia standard," in *Proc. IEEE International Caracas Conference on Devices, Circuits and Systems,* pp. 153–158, March 1998.

[109] T. S. Rappaport, *Wireless Communications: Principles & Practice,* Prentice Hall, 1996.

[110] D. Reininger, R. Izmailov, B. Rajagopalan, M. Ott, and D. Raychaudhuri, "Soft QoS control in the WATMnet broadband wireless system," *IEEE Personal Communications Magazine,* vol. 6, no. 1, pp. 34–43, Feb. 1999.

[111] I. Rhee, "Error control techniques for interactive low-bit-rate video transmission over the Internet," in *Proc. ACM SIGCOMM'98,* Vancouver, Canada, Aug. 1998.

[112] G. G. Roussas, "A course in mathematical statistics," 2nd ed., Academic Press, 1997.

[113] A. Sampath, P. S. Kumar, and J. M. Holtzman, "Power control and resource management for a multimedia CDMA wireless system," in *Proc. IEEE PIMRC'95,* Sept. 1995.

[114] H. Schulzrinne, S. Casner, R. Frederick, and V. Jacobson, "RTP: a transport protocol for real-time applications," *RFC 1889,* Internet Engineering Task Force, Jan. 1996.

[115] C. E. Shannon, "A mathematical theory of communication," *Bell System Technical Journal,* vol. 27, 1948.

[116] S. Shenker, C. Partridge, and R. Guerin, "Specification of guaranteed quality of service," *RFC 2212,* Internet Engineering Task Force, Sept. 1997.

[117] Y. Shoham and A. Gersho, "Efficient bit allocation for an arbitrary set of quantizers," *IEEE Trans. on Acoust., Speech and Signal Processing,* vol. 36, no. 9, pp. 1445–1453, Sept. 1988.

[118] M. K. Simon and M.-S. Alouini, "Digital communication over fading channels," Wiley, 2000.

[119] B. Sklar, "Rayleigh fading channels in mobile digital communication systems Part I: characterization," *IEEE Communications Magazine,* vol. 35, no. 7, pp. 90–100, July 1997.

[120] H. M. Smith, M. W. Mutka, and E. Torng, "Bandwidth allocation for layered multicasted video," in *Proc. IEEE Int. Conference on Multimedia Computing and Systems,* vol. 1, pp. 232–237, June 1999.

[121] K. Sripanidkulchai and T. Chen, "Network-adaptive video coding and transmission," in *SPIE Proc. Visual Communications and Image Processing (VCIP'99),* San Jose, CA, Jan. 1999.

[122] W. Stallings, "Data & computer communications," 6th ed. pp. 208–233, Upper Saddle River, NJ: Prentice-Hall, 2000.

[123] G. J. Sullivan and T. Wiegand, "Rate-distortion optimization for video compression," *IEEE Signal Processing Magazine,* pp. 74–90, Nov. 1998.

[124] H. Sun and W. Kwok, "Concealment of damaged block transform coded images using projection onto convex sets," *IEEE Trans. on Image Processing,* vol. 4, pp. 470–477, Apr. 1995.

[125] H. Sun, W. Kwok, M. Chien, and C. H. J. Ju, "MPEG coding performance improvement by jointly optimizing coding mode decision and rate control," *IEEE Trans. on Circuits and Systems for Video Technology,* vol. 7, pp. 449–458, June 1997.

[126] R. Talluri, "Error-resilience video coding in the ISO MPEG-4 standard," *IEEE Communications Magazine,* pp. 112–119, June 1998.

[127] W. Tan and A. Zakhor, "Multicast transmission of scalable video using receiver-driven hierarchical FEC," in *Proc. Packet Video Workshop 1999*, New York, April 1999.

[128] W. Tan and A. Zakhor, "Real-time Internet video using error resilient scalable compression and TCP-friendly transport protocol," *IEEE Trans. on Multimedia*, vol. 1, no. 2, pp. 172–186, June 1999.

[129] T. Turletti and C. Huitema, "Videoconferencing on the Internet," *IEEE/ACM Trans. on Networking*, vol. 4, no. 3, pp. 340–351, June 1996.

[130] T. Turletti, S. Parisis, and J. Bolot, "Experiments with a layered transmission scheme over the Internet," *INRIA Technical Report*, http://www.inria.fr/RRRT/RR-3296.html, Nov. 1997.

[131] J. S. Turner, "New directions in communications (or which way to the information age?)," *IEEE Communications Magazine*, vol. 24, no. 10, pp. 8–15, Oct. 1986.

[132] B. Vandalore, R. Jain, S. Fahmy, and S. Dixit, "AQuaFWiN: adaptive QoS framework for multimedia in wireless networks and its comparison with other QoS frameworks," in *Proc. IEEE Conference on Local Computer Networks (LCN'99)*, Boston, MA, USA, Oct. 17–20, 1999.

[133] A. Vetro, H. Sun, and Y. Wang, "MPEG-4 rate control for multiple video objects," *IEEE Trans. on Circuits and Systems for Video Technology*, vol. 9, no. 1, pp. 186–199, Feb. 1999.

[134] L. Vicisano, L. Rizzo, and J. Crowcroft, "TCP-like congestion control for layered multicast data transfer," in *Proc. IEEE INFOCOM'98*, vol. 3, pp. 996–1003, March 1998.

[135] J. Villasenor, Y.-Q. Zhang, and J. Wen, "Robust video coding algorithms and systems," *Proceedings of the IEEE*, vol. 87, no. 10, pp. 1724–1733, Oct. 1999.

[136] A. J. Viterbi, "CDMA: principles of spread spectrum communication," Addison Wesley, 1995.

[137] H. Wang, "Opportunistic transmission for wireless data over fading channels under energy and delay constraints," *Ph.D. Dissertation*, Department of Electrical and Computer Engineering, Rutgers, The State University of New Jersey New Brunswick, New Jersey, Jan. 2003.

[138] J.-T. Wang and P.-C. Chang, "Error-propagation prevention technique for real-time video transmission over ATM networks," *IEEE Trans. on Circuits and Systems for Video Technology*, vol. 9, no. 3, April 1999.

[139] X. Wang and H. Schulzrinne, "Comparison of adaptive Internet multimedia applications," *IEICE Trans. on Communications,* vol. E82-B, no. 6, pp. 806–818, June 1999.

[140] Y. Wang and S. Lin, "A modified selective-repeat type-II hybrid ARQ system and its performance analysis," *IEEE Trans. Commun.,* vol. 31, pp. 593–608, May 1983.

[141] Y. Wang, Q.-F. Zhu, and L. Shaw, "Maximally smooth image recovery in transform coding," *IEEE Trans. on Communications,* vol. 41, no. 10, pp. 1544–1551, Oct. 1993.

[142] Y. Wang, M. T. Orchard, and A. R. Reibman, "Multiple description image coding for noisy channels by pairing transform coefficients," in *Proc. IEEE Workshop on Multimedia Signal Processing,* pp. 419–424, June 1997.

[143] Y. Wang and Q.-F. Zhu, "Error control and concealment for video communication: A review," *Proceedings of the IEEE,* vol. 86, no. 5, pp. 974–997, May 1998.

[144] T. Wiegand, M. Lightstone, D. Mukherjee, T. G. Campbell, and S. K. Mitra, "Rate-distortion optimized mode selection for very low bit-rate video coding and the emerging H.263 standard," *IEEE Trans. on Circuits and Systems for Video Technology,* vol. 6, pp. 182–190, April 1996.

[145] J. Wroclawski, "Specification of the controlled-load network element service," *RFC 2211,* Internet Engineering Task Force, Sept. 1997.

[146] T. Hou, D. Wu, B. Li, I. Ahmad, H. J. Chao, "A differentiated services architecture for multimedia streaming in next generation Internet," *Computer Networks,* vol. 32, no. 2, pp. 185–209, Feb. 2000.

[147] D. Wu and H. J. Chao, "Efficient bandwidth allocation and call admission control for VBR service using UPC parameters," *International Journal of Communication Systems,* vol. 13, no. 1, pp. 29–50, Feb. 2000.

[148] D. Wu, T. Hou, Z.-L. Zhang, H. J. Chao, "A framework of node architecture and traffic management algorithms for achieving QoS provisioning in integrated services networks," *International Journal of Parallel and Distributed Systems and Networks,* vol. 3, no. 2, pp. 64–81, May 2000.

[149] D. Wu, Y. T. Hou, B. Li, W. Zhu, Y.-Q. Zhang, and H. J. Chao, "An end-to-end approach for optimal mode selection in Internet video communication: theory and application," *IEEE Journal on Selected Areas in Communications,* vol. 18, no. 6, pp. 977–995, June 2000.

[150] D. Wu, Y. T. Hou, W. Zhu, H.-J. Lee, T. Chiang, Y.-Q. Zhang, and H. J. Chao, "On end-to-end architecture for transporting MPEG-4 video over the Internet," *IEEE Trans. on Circuits and Systems for Video Technology*, vol. 10, no. 6, pp. 923–941, Sept. 2000.

[151] D. Wu, Y. T. Hou, and Y.-Q. Zhang, "Transporting real-time video over the Internet: challenges and approaches," *Proceedings of the IEEE*, vol. 88, no. 12, Dec. 2000.

[152] D. Wu, Y. T. Hou, and Y.-Q. Zhang, "Scalable video soding and transport over broadband wireless networks," *Proceedings of the IEEE*, vol. 89, no. 1, Jan. 2001.

[153] D. Wu, T. Hou, W. Zhu, Y.-Q. Zhang, J. Peha, "Streaming video over the Internet: approaches and directions," *IEEE Transactions on Circuits and Systems for Video Technology*, Special Issue on Streaming Video, vol. 11, no. 3, pp. 282–300, March 2001.

[154] D. Wu, T. Hou, H. J. Chao, B. Li, "A per flow based node architecture for integrated services packet networks," *Telecommunication Systems*, vol. 17, nos. 1,2, pp. 135–160, May/June 2001.

[155] D. Wu and H. J. Chao, "Buffer management and scheduling schemes for TCP/IP over ATM-GFR," *International Journal of Communication Systems*, vol. 14, no. 4, pp. 345–359, May 2001.

[156] D. Wu and R. Negi, "Effective capacity: a wireless link model for support of quality of service," *IEEE Trans. on Wireless Communications*, vol. 2, no. 4, pp. 630–643, July 2003.

[157] D. Wu and R. Negi, "Downlink scheduling in a cellular network for quality of service assurance," *Technical Report*, Carnegie Mellon University, July 2001.

[158] D. Wu and R. Negi, "Utilizing multiuser diversity for efficient support of quality of service over a fading channel," *Technical Report*, Carnegie Mellon University, Aug. 2002.

[159] X. R. Xu, A. C. Myers, H. Zhang, and R. Yavatkar, "Resilient multicast support for continuous-media applications," in *Proc. IEEE Int. Workshop on Network and Operating System Support for Digital Audio and Video (NOSSDAV'97)*, pp. 183–194, May 1997.

[160] O. Yaron and M. Sidi, "Performance and stability of communication networks via robust exponential bounds," *IEEE/ACM Trans. on Networking*, vol. 1, no. 3, pp. 372–385, June 1993.

[161] D. Zhang and K. Wasserman, "Transmission schemes for time-varying wireless channels with partial state observations," in *Proc. IEEE INFOCOM'02*, June 2002.

[162] H. Zhang, "Service disciplines for guaranteed performance service in packet-switching networks," *Proceedings of the IEEE*, vol. 83, no. 10, Oct. 1995.

[163] L. Zhang, S. Deering, D. Estrin, S. Shenker, and D. Zappala, "RSVP: A new resource ReSerVation Protocol," *IEEE Network Magazine*, vol. 7, no, 5, pp. 8–18, Sept. 1993.

[164] Q. Zhang and S. A. Kassam, "Hybrid ARQ with selective combining for fading channels," *IEEE Journal on Selected Areas in Communications*, vol. 17, no. 5, pp. 867–880, May 1999.

[165] Q. Zhang and S. A. Kassam, "Finite-state markov model for Rayleigh fading channels," *IEEE Trans. Commun.*, vol. 47, no. 11, pp. 1688–1692, Nov. 1999.

[166] R. Zhang, S. L. Regunathan, and K. Rose, "Video coding with optimal inter/intra-mode switching for packet loss resilience," *IEEE Journal on Selected Areas in Communications*, vol. 18, no. 6, pp. 966–976, June 2000.

[167] Z.-L. Zhang, "End-to-end support for statistical quality-of-service guarantees in multimedia networks," *Ph.D. Dissertation*, Department of Computer Science, University of Massachusetts, Feb. 1997.

[168] Z.-L. Zhang, S. Nelakuditi, R. Aggarwa, and R. P. Tsang, "Efficient server selective frame discard algorithms for stored video delivery over resource constrained networks," in *Proc. IEEE INFOCOM'99*, pp. 472–479, New York, March 1999.

[169] Z.-L. Zhang, Z. Duan, and Y. T. Hou, "Virtual time reference system: a unifying scheduling framework for scalable support of guaranteed services," *IEEE Journal on Selected Areas in Communications*, vol. 18, no. 12, pp. 2684–2695, Dec. 2000.

[170] M. Zorzi, R. R. Rao, and L. B. Milstein, "Error statistics in data transmission over fading channels," *IEEE Trans. Commun.*, vol. 46, no. 11, pp. 1468–1477, Nov. 1998.

[171] M. Zorzi, A. Chockalingam, and R. R. Rao, "Throughput analysis of TCP on channels with memory," *IEEE Journal on Selected Areas in Communications*, vol. 18, no. 7, pp. 1289–1300, July 2000.

Wissenschaftlicher Buchverlag bietet

kostenfreie

Publikation

von

wissenschaftlichen Arbeiten

Diplomarbeiten, Magisterarbeiten, Master und Bachelor Theses
sowie Dissertationen, Habilitationen und wissenschaftliche Monographien

Sie verfügen über eine wissenschaftliche Abschlußarbeit zu aktuellen oder zeitlosen
Fragestellungen, die hohen inhaltlichen und formalen Ansprüchen genügt,
und haben **Interesse an einer honorarvergüteten Publikation**?

Dann senden Sie bitte erste Informationen über Ihre Arbeit per Email
an info@vdm-verlag.de. Unser Außenlektorat meldet sich umgehend bei Ihnen.

VDM Verlag Dr. Müller Aktiengesellschaft & Co. KG
Dudweiler Landstraße 125a
D - 66123 Saarbrücken

www.vdm-verlag.de

www.ingramcontent.com/pod-product-compliance
Lightning Source LLC
LaVergne TN
LVHW022304060326
832902LV00020B/3256